Christopher Columbus Answers All Charges

To Sheila
With best wishes,

Voyage on!
Marc

Christopher Columbus Answers All Charges

Yuri Rubinsky and Marc Giacomelli

The Porcupine's Quill, Inc.

CANADIAN CATALOGUING IN PUBLICATION DATA

Rubinsky, Yuri
  Christopher Columbus answers all charges

ISBN 0-88984-150-0

1. Columbus, Christopher, ca. 1451-1506 – Fiction.
1. Giacomelli, Marc. 11. Title.

PS8585.U24C5 1993   C813'.54   C92-094952-5
PR9199.3.R83C5 1993

Published by The Porcupine's Quill, Inc., 68 Main Street, Erin, Ontario NOB 1TO with financial assistance from The Canada Council and the Ontario Arts Council. The support of the Government of Ontario through the Ministry of Culture and Communications is also gratefully acknowledged.

Cover is after a photograph of a parking lot in Columbus, Ohio, taken by Tim Kibbey. Electronic image enhancement by Carole Bourgeois.

Distributed by General Publishing, 30 Lesmill Road, Don Mills, Ontario M3B 2T6.

Thank you, immeasurably, to Holley and Sarah and to our families for their stories in these stories. To Stan Bevington for technological support and encouragement, Kate Hamilton for original typesetting, Michelle Dunne, Eva Rennau, Doris Cowan for copyediting, and the people at The Porcupine's Quill. To Richard Landon at the Thomas Fisher Rare Books Room at the University of Toronto. Plus all the supportive people at SoftQuad, Robins Sharpe and the Queen's Own Rifles of Canada. To the Ontario Arts Council, and to Louise Dennys, Catherine Yolles, Eli Robinsky and Paolo Friedenberg for their very helpful suggestions as first readers.

Contents

CRIDCRIDCRIDCRIDCRIDCRIDCRIDCRIDCRIDCRIDCRID

A Chronology of the Life of the Admiral Christopher Columbus as is Best Known Hereto.

1446 – The Admiral is born in the Republic of Genoa. Some time later he is apprenticed by his father as a wool carder, combing and cleaning raw wool.

1460 – The Admiral begins his life as a mariner.

1474 – The Admiral corresponds with Paolo Toscanelli, cosmographer of Florence, and begins his studies of *The Travels of Marco Polo* and *The Imago Mundi* of Petrus Aliacus.

1474 – Travels of the Admiral to England, Iceland, and other places.

1478 – Marriage of the Admiral in Lisbon to Felipe Moniz, daughter of a captain of Prince Henry the Navigator of Portugal.

1481 – Earliest record of compilation by the Admiral of the prophecies which will eventually become part of his so-called *Book of Prophecies*.

1482 – Time during which the Admiral, travelling to Porto Santo and the Gold Coast, develops his theories of the world as a sphere and the project of a Western enterprise to reach the Indies.

1484 – The Admiral's theories are rejected by the Council for Geographical Affairs of King John of Portugal.

1487 – Theories rejected by the Spanish Sovereigns at the so-called Dominican Commission of Salamanca.

1488 – Second rejection of the theories by the King of Portugal.

1490 – The Admiral's plans, as represented by his brother, Bartholomew, rejected by Henry VII of England.

1491 – Second Spanish rejection by the Commission of Cardinal Mendoza at Granada.

1492 – The brother of the Admiral rejected by the Court of France and the capitulation of Ferdinand and Isabella to the demands of the Admiral.

1492 – Friday, August 3rd, the Admiral's three ships leave Palos, Spain for the New World, where they land Friday, October 12th, at San Salvador. Founding of the colony at La Navidad, Hispaniola and return of the Admiral, after many delays, to Palos in March of the following year.

1493 – The second voyage of the Admiral in seventeen ships in September. La Navidad is discovered burned. Other colonies are founded and after much concern with the natives, and the desertion of appointed officials, the Admiral leaves his brother as Adelantado and returns to Cadiz in 1496 dressed in sackcloth.

1498 – The third voyage, during which Trinidad is named, the great rivers called the Mouth of the Dragon which lead to the Garden of Eden are seen, and the continent touched upon. The Admiral calms the natives and settlers at Hispaniola, but is delivered with his brothers in chains to Francisco Bobadilla, Royal Judge, sent by schemers at the Court.

1500 – The Admiral, manacled, returns to the Sovereigns, is received with honour and distinction and the charges repudiated.

1501 – First collection of *Notebook of authorities, statements, opinions and prophecies on the subject of the recovery of God's holy city and mountain of Zion, and on the discovery and evangelization of the islands of the Indies and of all other peoples and nations*, later called the *Book of Prophecies*.

1502 – The fourth voyage of the Admiral, during which there is a great tempest at Hispaniola, an isthmus is discovered, and tales are heard of a rich land and another ocean to the East. The Admiral suffers much while stranded at Yemyaica for a year. He returns to Sevilla in 1504.

1505 – The Admiral removes himself to the Court at Segovia and thence, alone, to Valladolid.

1506 – The Admiral, now in his sixtieth year, begins writing the present work, called by him *Contra Impetus Omnes*.

A Few Words of Preface

Reader: There is a present time for this book also, distinct from the time recollected throughout its pages.

As I dictate these passages, I sit in a small courtyard of the monastery of Santa Maria de Las Cuevas in Sevilla. I am surrounded by plantings of small red flowers – I who have named lands and seas cannot remember the name of this flower; if I were younger I would give it a new name – and I take pleasure from the orderliness of the rock pathways that divide this court into rectangles, one with a fountain, one with close-cropped grasses, many with patterns of these small red flowers.

I am old. Voyages ago my red beard turned to white, but now my face too seems that of a wise, worn explorer. I look successful; some would argue, misleadingly so.

You, reader, who live 50 or 100 years hence, I can only know so much about you. I cannot know whether you pay tribute to the Grand Turk or the Great Khan, or whether there is a Supreme Pontiff in some New Rome. You may live in a Jerusalem free at last or in the Christian Jipangu or on one of the islands of tragedy so near to my heart. You may make obeisance to one of my family (may they live on), as Vice Regent.

But I know what is most important: I know you are living in the Final Age, watching God's plan for history come to fruition as a result of my actions. I know you are seeing evidence all around you of a great Christian colonizing of the world, and the other signs of the Second Coming of Christ and the End of the World.

A man who pulls himself to a great success, and then falls out of favour with those who should most appreciate his accomplishments, such a man gathers moments around him like courtiers, and relives every moment, and every accusation, and every failure. But such a man explores also the after-effects and the tangential consequences of his discoveries, imagining as best the human kind can (within the limits of the knowledge of our fate established by the grace of Providence) where events must or may follow other events, in which action a seed was planted of larger, later actions, and how those seeds should grow, wither, or flourish.

This too I have done. I can know the future without seeing it. And I can understand my place in its foundation, which is, naturally, my day, our days.

These few pages are my notes on my voyage through life, a life filled with faith, hope and discovery. (Is it true that anyone writing of his life over-uses metaphors taken from his own calling? Do I write with images of discovery to such a point that each reader seeks one quiet, friendly page free of allusion to exploration and the sea?)

So be it. Each of us lives in an age of discovery without maps, without instruments, without end. (Something can be a metaphor of the sea and also be true.)

I am intrigued by this courtyard because it is walled and it contains pathways. Looking back on my life, I see that I have spent more time in this humble courtyard than anywhere else with walls, and at no time in my life have I had such literal paths to follow.

I sit on a stone bench. As I speak, my friend Father Gorricio, or sometimes my son Ferdinand, writes in my notebook. First I or we recollect one of the many charges that have been made against me; I become angered anew, or amused, or wistful; and then I compose a response which I dictate. A slow process, and at times a painful one, but I find myself surprisingly at ease between these stone walls and walking slowly along these stone paths, polished to smooth by the soft shoes of holy men who have paced here before me.

I dedicate this book to the Holy Mother and the Blessèd Saints, to the memory of Her Majesty the Queen Isabella, to the memories of the mothers of my sons, and to the spirit of my mother, which has continued to guide me through each of my voyages and each of these pages.

In my advanced years, I have a tendency to confuse these women, and to forget which is holy, which is a spirit, which still lives.

Accusations General in Nature

That I knew not whither I sailed.

On a late afternoon in August, 1492, seven days out from the Canarias, four of us stood on the aftcastle deck, staring at the sea.

Rodrigo de Escobedo pointed at what could have been a dark patch of weeds and shouted, 'There, a leviathan! It has tusks like a narwhal, and three spouts. It forces water in the air taller than our masts, in nine directions at once.'

'Look, on the other side,' said Luis de Torres, the translator, 'there be dragons. The same creature described by the ancients that rescued the lords of Atlantis and carried them to safety on its back.'

We were amusing ourselves by making gentle mockery of the ornaments drawn by map-makers to break the monotony of any vast expanse of ocean. My brother Bartolomeo, a cartographer by training, was standing at his drawing platform and we were gathered about him. He had drawn our small ships on a sheet of vellum and was now surrounding us with each monster in turn as it was described.

The others having spoken, they turned to me.

'There,' I said pompously, pointing towards the horizon directly ahead of us, 'most terrifying: the edge of the map.'

That I am vain beyond all reason, beyond that pride which one would naturally expect in a man of my accomplishments.

What does it mean to be first? To undertake an action in God's name, and be successful beyond the expectation of one's colleagues and countrymen. To become the victim of envy and small jealousies, of false scandals and incredible lies. To be heralded as a Saint, and then feel oneself a Martyr. To feel sometimes as Moses must have in his solitude, sometimes like the Blessèd Catherine.

I remember the night that I feared I had wearied of the sea.

The sun was low in the sky, we were maintaining a reasonable speed, the ocean swells were those that rock a tired man to sleep but keep a slightly alert sailor awake.

We had been seeing the signs of land for several days – certain birds, tree branches, a log whose limbs had been unmistakably removed by men.

This was my third crossing. Our fleet was of six ships, and we were closer to the equatorial line, indeed closer to the south and west points of the charted seas than ever sailors had been. It had been known for many centuries that as one approaches the equator, in climates more and more torrid, the works of nature are more abundant, more perfect and more precious.

Naturally, my thoughts turned to the lands we would see the next day, and the discoveries we would make, the precious stones, the gold, the spices, the powders.

In the next instant, I thought too of the claims that men would make against me, of the inevitability that whatever proofs I collected and presented to the court, someone would dispute their origin or their authenticity or their value.

Suddenly, I was greatly tired. I became physically ill. Each roll of the ship seemed to have an effect on my stomach, a sensation I had never before experienced. Soon I recognized that the sea was acting towards me just as my enemies would in Spain and Portugal when I returned, that I was being tossed relentlessly, that I was fragile, and perhaps defenceless, and above all, too tired to struggle, either with the sea, whose every wave now had to be conquered, or with my enemies on land.

The bounds of the sea I recognized and understood. I didn't know even the names of my enemies. Indeed, some, on that day in 1498, probably still counted themselves among my friends.

Today I recognize that some of my enemies are not yet born.

What does it mean to be first? To have invisible enemies and enemies beyond your deserving.

Were I a lesser man, or were my accomplishments less important, my enemies and their accusations would die before me.

I am strong but I will die.

My children will defend me against my accusers, and their children, and theirs. They will inherit my governance and my property, such as it is – its paucity is a sign of the strength of my enemies – and they will inherit a place in a world changed by their father.

Naturally, I grieve for my children. They will inherit the injustice that has befallen me, and because they are clever, and they have sailed with me, and know me at my strongest, they will inherit my tenacity, and my fervent desire for our justice.

Yes, I am vain beyond all reason. I am pursued beyond all reason, I am mistreated beyond all reason, I will persevere beyond all reason.

That I was misled by Paolo del Pozzi Toscanelli into my geographic beliefs the evils of which continue to this day.

I corresponded with the eminent Toscanelli, but he was only one of the sources of my knowledge and certainty.

Toscanelli was right in principle and wrong in detail. Who would not be charmed by his tales of a world that was round and met itself on the other side? Who would not be comforted by his maps that show all known lands in their relationships, each to each? And that show there are not great spaces of the sea to be traversed which are unknown, and cannot be measured? The elegance of geographic speculation leaves a trail of misnamed lands and misunderstood natives and spectral khans.

But do your children not believe in a moon that is the head of a white-masked Venetian wrapped in a cloak and a sun that sleeps each night? And are not even now mapmakers creating other undiscovered worlds of coasts and stars and monsters in the blackness?

Some read Toscanelli, some do not. Some are offended by him. Some will wait for his successors. Nonetheless, discoverers will go as I went: exhorting a vision, pleading for money, promising new lands of wealth and freedom and a new beginning. They too one day will stand accused and you with them; for glorifying their exploits, pointing them out in the streets as heroes to your children, for believing in their very discoveries.

But your tragedy and the tragedy of those you discover will be worse than mine, no matter how true your maps and methods, if you go without God.

It is a sign of our age, an age of moral and physical cowardice, that understanding Toscanelli and the many other sources is not the critical step in beginning to explore. Success in pleading for money is the sign of a true explorer today.

I recognize a certain bitterness in this point of view.

I remind you of the story of the egg.

Upon my return from the first voyage, I was greeted with great adulation, and named Admiral by Their Majesties. Pedro Gonzalez de Mendoza, Grand Cardinal of the Spanish Empire, a man of great wisdom, piety and position, invited me to dine with him, and accorded me such honours as he might have bestowed upon King Ferdinand himself.

A jealous person, of no value or station, accusing me of being a foreigner (I who had served Their Spanish Majesties faithfully and vigorously), asked me if there were not other men in Spain who could have accomplished the discovery of the Indies.

I did not reply, but took an egg from the plate in front of me, and passed it to the man beside, inviting him and all there present to make this egg stand on its end.

Each guest tried this test, and the egg was passed on to the next, each as unsuccessful as the one before. After some time, the egg was passed back to me. I held it in one hand and struck it hard, on one end, so as to break the shell. I left it standing on its end on the table.

That I prayed for the deaths of my enemies.

The collection of events surrounding the fourth voyage will form another part of this narrative.

I have generally found that the death of an enemy, while a blessing, should not be considered a personal favour from the Almighty.

That I foolishly and boldly demanded the title 'Admiral of the Ocean Sea' when no such title existed before me.

Who has gone before me that deserved such rank? And who has achieved such rank before his ever sailing? This should convince my accusers as I convinced Their Most Catholic Majesties by my arguments, charts and plans that I am deserving not only of the title but of the benefices and emoluments attached thereto. Namely: one-eighth of the revenues as a whole from new territorial rights and jurisdictions; one-tenth the revenue once charged with costs and expenses, and one-third thereafter.

And with this title, that of Viceroy of new lands touched upon by the Discoverer. In addition, the Royal Patent commanding all prelates and masters, counts, grandees, knights and esquires, counsellors, alcades, bailiffs, justices, prestameros, provosts and other officials of all the cities and towns of the kingdoms and dominions, and captains of the sea, superintendents of the fleet, masters and boatswains of the galleys and the masters, mariners, merchants and other persons who traverse and navigate the ocean seas to receive and obey Don Cristobal Colon, said Admiral.

Further, none but I had heretofore been furnished with Royal Letters Introductory to the Grand Khan of Tartary, the Emperor of Jipangu and other such potentates as I should meet upon my travels.

All this was granted before I set one sail aloft in the service of the Spanish Sovereigns. This is surely evidence enough that rank and title need not only follow deeds but can command them of one ennobled by his office.

By choosing a title that had never before existed, I imagined that I would avoid accusations such as this one.

That I allowed a stowaway girl to sail with me
on the Santa Maria.

In Sevilla, before the Queen had determined to support me in my mission, I was approached by a servant of the Queen's ten-year-old daughter, Johanna, who is now, may God preserve her, Queen of Castile, and not of sound mind.

This servant, self-possessed and educated beyond her position, introduced herself as Susana Fontanarossa. My astonishment was great. She had the red hair of my mother, and the same name. She could not have been more than sixteen years old and she told me that she once had a dream that she would sail to new lands and that in this dream she would try to communicate with strange creatures who spoke no language at all.

Whether it was her name, or her familiar look, or her determination, or my curiosity regarding this dream, I chose to invite her to relate it.

'I commanded a ship,' she said, 'as Johanna d'Arc la Romée commanded an army.'

I interrupted her. 'I recognize that the Pope and the Bishops have annulled the charges against this girl, but no one has now said that girls of eighteen should lead armies.'

'In my dream, I commanded a ship,' she repeated, 'one of three ships which sailed west across the Ocean Sea. You were on another. I dressed as a man does, and remained a virgin, and spoke with the blessèd Saints Catherine and Margaret, as Johanna did.

'After some time, we arrived at a mountain, and there were people living at its base who had only one eye and had green hair. When you or any of your men tried to speak with them, they made noises like animals. I asked you to let me try to communicate with them, and you agreed.

'But before I had a chance to take part, you said, "We know nothing of this way of life," and you took your hat and threw it angrily on the ground. One of the creatures fashioned a hat out of some branches and placed it at his feet. Then he walked to your hat, and turned it over. Then back to his hat, and turned it over too.

'One of your men unfurled a flag, and immediately one of the creatures unfurled a flag, and then killed the sailor. Then this creature was killed by his comrades.

'You pointed to yourself and waved towards the ships in the harbour. "Ships," you said. "We have sailed here from across the Sea."

'One of the creatures stepped forward and before any of us could say or do anything, he ate your hat. You had parchment in your hand, and signed your name. Another of the creatures suddenly sat down. You picked up a leaf and folded it in half. One of the creatures pointed at the ships, which at that moment disappeared.

'Finally I spoke. I said that the Saints have commanded that we teach these beings the idea of singing, and so I sang a hymn that I learned from nuns. And they stood quietly and listened, motionless. Then I said that what we needed was a way of telling these creatures that sometimes we were trying to talk with them, and that sometimes we were offering them gifts, and sometimes we were declaring war. We needed to have a language to tell them what language we would speak to them, a separate language whose only role was to warn them that what would next befall them was communication, or food, or war. Spanish, I said, would not be helpful to us in this situation unless we told them first that we would be speaking Spanish.

' "But in what language could we tell them this?" I asked, and you did not reply. I woke up.'

This person did not sail with us on the *Santa Maria*. There were no stowaways on that voyage.

I do not understand the dreams of others. I have very simple dreams. I do not understand how someone who is not from Istria could have the name of Fontanarossa. When I think of this girl now, fifteen years later, I think again of my mother, and when I think of my mother, I think of this girl.

It disturbs me sometimes that I distinguish between dreams and prophecies. Later, when this servant girl sailed with us, she tried to convince me that prophets were people who wrote down their dreams, and that if she knew how to write, she would be a prophet too.

I still, however, cannot explain her dream. She must have simply guessed that I would sail with three ships.

That I did little to protect the natives of the Indies
from mistreatment by the Spanish.

In 1502, one month at sea from Spain, we sat on the chairs from my cabin on the aftcastle deck in mid-morning. Palahuelo, who was also called Kee, an Indian shaman, which is their notion of doctor and priest together, was returning with me from Spain where he had lived for two years with Franciscan monks. I try here, from my weak memory and scant Journal notes, to reconstruct the conversation.

'There are only two types of Spaniard,' Kee said, 'and you are in the second category. The first thinks we are savages, barely more useful than fish, incapable of reason, perhaps incapable of feeling pain when we are hurt, feeling sorrow when we are taken from our families. Or, as in my case, leaving my family willingly, to be an explorer and discoverer, sailing off the edge of the horizon to Castile.

'The second kind of Spaniard thinks we are savages, barely more capable of conversion to Christianity than Jews or Mohammedans, but just enough likely to convert that we are worth keeping alive. But these Spaniards also think we are creatures of beauty, reflecting God's grace in our docility, our handsome features, even in our nakedness.

'You, my friend,' he continued, 'you are like the Queen, or Father Perez. I believe you think we are a gift from God to you. You mean us well, you believe the Christian faith will give us strength and comfort as it does you. Certain of you have indicated in your writings and your speech that we have beliefs of our own, for which you have respect. This is an improvement over others who think we are as empty vessels into which they can pour Christianity as if it were water or wine.'

'Except that I am not a Spaniard, your observations, as always, are accurate,' I said, 'but I hope that you see us as intending the best for you, with our religion, and our missionaries, and our books and our writings.'

'Sometimes I think that you care very little as to the depths of our understanding and commitment to the Christian faith, so long as you can mark on your paper that you have preached to yet another collection of heathens, that you have fulfilled yet another ancient prophecy.'

'This is not true!' I remonstrated. 'You cannot know the depths of our concern for your spiritual well-being.'

'I think the opposite,' said Kee. 'It is you who cannot measure the depths of your own concern for us. We see these depths in your eyes, in your trinkets, and in your swords.'

That I was secretly a Jew of Venice, disguised my apostasy, and escaped the Inquisition to the new world.

I remind you that Jesus Christ our Lord was born a Jew. And resurrected a Catholic. I am neither morisco, marrano nor converso, but a servant, as often said, of Our Lord and His Most Catholic Majesties. This story has been put about by my enemies out of jealousy for the fact that, having been rejected by the Sovereigns of Europe, including the Queen and King of Spain, I was disconsolately leaving Granada to continue my explorations alone and on land when a rider from Don Luis de Santangel, Receiver of Ecclesiastical Revenues for the Crown of Aragon and himself secretly a Jew, stopped me on the bridge at Pinos and throwing a pebble into the stream said, 'Sail if you must, your demands have been granted.'

Thus did Their Majesties capitulate to my arguments and terms: named Admiral of the Ocean Sea, Vice-Regent of all I should discover and deserving of a greater part than they of all precious metals within my Admiralty. The terms reminded my enemies, who are many, of those of certain Jewish tax collectors and money lenders.

But if Their Majesties had thought me a Jew, I assure you that they know there are simpler ways of ridding oneself of Jews than paying them to go.

That the divine Providence, in telling us, through His Holy
Word, to preach that Word unto all the corners of the World,
intended for us to preach to the Muslim and the Jew in order that
they might follow the Teachings of Jesus and that the final days
might come as foreseen in the divine revelation of Saint John.
That He never intended us to seek other and new infidels in lands
not known during Jesus' lifetime on earth, and that to do so is a
vanity and a temptation and an interpretation of God's Will
made flexible to justify our vanities.

This accusation, although it has been directed at me by certain
priests and their followers, represents rather the view of those
voices who rail against the temper of their times no matter when
they live or what those times.

Likewise accused by those statements are the Royal Houses of
Spain, Portugal, England, Holland and Venice, all infused with
the role God has given them in discovering the undiscovered corners of His world. Similarly accused, then, are the Portuguese Da
Gama, and Diaz, the Venetian Caboto and countless others.

None of us undertakes a voyage alone or without backers.
Similarly accused, then, are the bankers and merchants who pay
the investment in our voyages and who look to profit by our success; the lawyers who determine the division of that success; the
fathers who send their sons to sea with us; the gaolers who send
us their prisoners; the ship-builders, the sail-makers, the provi-
sioners, the coopers, the harbour-masters who extract the taxes
and fees.

I suggest to you further that if we who discover in the Name
of God and our Kings and Queens are to be accused of vanities
then similarly accused must be all those men and women who
surround us and our ships when we sail and who run to the harbour to greet us on our return.

What of those infidel strangers that populate the Indies?
I confess that their presence and their manner surprised me.

From my study of the works of many scholars, I know that the subjects of the Khan are learned, they speak several languages, and that they are able to debate with us on the subject of our and their beliefs. M. Polo tells us they have thousands of ships in their harbours in the Ocean Sea. These are the harbours into which we expected to sail. I brought priests and lawyers that we might enjoy learned and complex discourse.

The Indians whom we met knew, naturally, of the empire of the Khan, and indicated to us that it was to be found with just a few days' sailing to the West and to the South. I believe that their inexperience on the sea and their lack of allegiance to or interest in the Khan made them underestimate that distance. They indicated also that there was much gold in that land, but expressed little interest in that most holy of metals. I believe that they answered our questions without guile and without knowing that they did not answer well.

At first, of course, our interest in these island people was not great, except insofar as they could direct us to the empire of the Khan and the riches thereof.

This was a kind of blindness, and we came to realize that there was much good that we could impart to them. We gave them seeds and seedlings and taught them to clear lands and grow certain plants. Several we taught to read the Word of God, in which they had much interest, calling to their friends and recounting to them with animated expressions and voices the lessons of God.

I regret still that in establishing a fortress at La Navidad I left behind men of little conscience and given to licentious behaviour, who took the Indian women for their own and who held little regard for the lives of men or women. It pains me greatly to recount this, but I believe that in all likelihood the Indians of Española were justified in destroying our fortress and in assassinating many of the men within.

Wondrous it was that on my return, I was greeted warmly by these same Indians. Their consciences weighed on them, however, and rather than tell me the truth of the events of 1493, they fled. They could not bring themselves to speak ill of my men, although their provocation must have been great to react as they did. I found them to be kind and gentle always.

This accusation is one of vainness, and yes, it is a vanity that we came to these gentle people and showed them how to grow the plants of Christendom on their green islands. But they learned willingly and ate of the food we offered.

We gave them our diseases and they us theirs, and our weakest died, and both our races grew stronger accordingly.

They fought us on occasion, and they killed us and we them. And both sides grew sadder and then we learned to avoid the actions that would provoke the other. All this happened in less than ten years.

God set these people here, close to the Garden of Eden, that we might find them on our way to Cathay, that they might teach us and might learn from us. We have not yet learned from them all that they might teach, but what we teach, we teach generously, on behalf of all of you in Christendom who pray for us and our success and who recognize God's Will in all our actions.

*That I am given to my imagination and have suffered
for its vividness.*

I have anchored ships in most harbours and watched the land from the perfect distance.

I have anchored by night off cities teeming with secret lives made slightly less secret by lights, and sounds, and smells.

I have anchored by night off dark uninhabited islands, with sounds and smells and no lights.

And I have anchored by night off dark shapes and shadows not knowing if they secretly teem with lives I can neither see nor know.

In such an anchorage, I have glimpsed golden reflections off cliff-faces of a sunset behind me and sometimes thought I saw small fires.

In such an anchorage, close to a dark shore, I saw no flames but smelled smoke and burning flesh and heard no sounds.

That I commit sacrilege with my assertions about my role in the history of Christendom.

I have said that only Jesus had a greater role in the history of Our Holy Church than I. I have explained that my part was a simple one, but one which was foreseen in prophecy.

Jesus encouraged His Saints to go to all the corners of the world, that all men might hear His Word and that they might all be brought into Christendom. This was one of the conditions to be fulfilled before He comes again.

Further, the prophecies indicate that brave men in the service of a Holy King and Queen shall unite all people in the faith of Christ in order that His Second Coming may be fulfilled. It is clear from the writings of Joachim of Fiore and other wise and holy prophets that it is our Gracious Majesties of Castile and of Spain to whom these writings refer.

Knowing this, I endured the taunts of their fops and the delays of their court in pressing for their support of my first voyage. I recognized that had I sailed in the service of another nation, I would not have been doing so in accordance with the learned writings and with the wishes of God.

I have learned, and I give this advice freely to all of you who may one day read these words, that if you have occasion to choose between two actions, and one of those two actions is in fulfilment of a prophecy, and one is not, though the way be far more difficult, do not hesitate in choosing the path wherein you fulfil the prophecies.

There is no sacrilege in my position. On my fourth voyage, I saw the Garden of Eden from whence sprang Eve and Adam and all of us.

I have sailed into the mouths of two of its four rivers. God has blessed me by showing me these wonders.

I have seen gold adorning the necks and hair of men and

women who had never heard the Word of God. The priests that I brought to the Indies showed these men and women the path to behaviour befitting the followers of Christ, and I was rewarded when these men and women described the origins of their gold and it was clear that this was gold from the mines of King Solomon. Now it is equally clear that gold from these mines will finance the raising of a Holy Army to capture Jerusalem in the name of all Christendom.

And who more fitting than I to lead such an army? In me, and through me, prophecies have been fulfilled. They shall be again.

*Accusations Concerning My Origins,
Youth and Family; Also Charges Arising
out of My Years of Preparation*

CRACRACRACRACRACRACRACRACRACRACRA

*That I am of low birth and that I have purposely made obscure
my origins and the history of my family.*

Let this speculation end. I was born in Genoa at the Porta
Sant'Andrea in the year 1446. As a child I wandered amongst
strange ships on the Mole Vecchio and listened to strange tongues
talking of the terrors beyond the seas.

Each day my mother would take me to the churches, each
built with the black and white marble that I came to associate
with all opportunities and all choices. Santo Stefano, home of the
first martyr, a Jew stoned to his death for calling the other Jews
stiff-necked. San Matteo, the church of commerce of the D'Oria
family. San Lorenzo, where lies displayed the Sacro Catino, the
glass and emerald cup of Our Lord's final supper, brought from
Caesarea, object of vain worship. Here was the birthplace of my
vision of a New Crusade financed by the as yet unattainable riches
of the lands beyond Jipangu. San Siro, Church of the Apostles,
where I watched odd men in odd hats elected as the Doges of
Genoa.

I am the son and the grandson of wool-combers and, but for
my recognizing, as a youth of unusual intelligence, that knowledge comes from only four sources, I would be such a wool-
comber also.

The four sources of knowledge are books, travel, discourse
with the learned, and revelation. Of these, none but books are
available to a young man embraced by the Guild of Wool-
combers.

I make no pretence to inherited titles. I have earned what I
have.

That I recognize too late that there is a danger in the
interpretation and misinterpretation of signs.

In the midst of a late afternoon storm off the south coast of Española, Susana Fontanarossa remarked that the tempest was a sign from God that we must go home.

I urged her not to talk of signs from God, saying that for two or three years in my youth, I became convinced that all events, all accidents of good and bad fortune, and all chance remarks were signs from God, sent to help me choose a path for my life.

At the time, I was apprenticed in the wool-combing trade. One day I was given a clump of wool to clean and card that had a black streak in it that looked very much like the outline of the island of Sardinia. 'A sign,' I thought, 'from God, to help me recognize my ability to find the physical reality in such an abstraction as black lines, a sign that I am to become a mapmaker.' Later I thought it was a sign that I was to become a wool-trader, and to build a home and an enterprise in Sardinia.

My brother was not swayed by my conviction.

'A sign from God,' he said, 'that the wool needs washing.'

Susana Fontanarossa, upon hearing my story, said that when she was young, her mother would frequently sing to her. When they went to church, there would be magnificent singing. Sometimes they would visit or be visited by cousins, who would sing for her too.

'Eventually,' she said, 'inevitably, as I walked through the market, still a very small child, amidst fresh fish and chickens and piles of vegetables, I heard singing coming from an upstairs window, a woman's voice, lower than my mother's. I stood transfixed, staring at the window, expecting the woman to come to wave at me and smile or call my name.

'Nothing. No one. The singing continued but no one acknowledged me. Passersby jostled by and stepped around. I caught up to my mother ahead, and was crying uncontrollably.

Eventually I became calm and tried to ask the right question and my mother explained that often people sing when they are alone, or with friends.

'She never understood the revelation that had caused my tears. I had thought that all singing was for me.'

'In this we are different,' I said. 'I have never completely out-grown the urge to recognize in all things God's instructions to me.'

Bartolomeo, upon reading these passages, insists that it was our own mother who told the story of the singing. He says that in my age and vigour, I have become confused.

That I lied about my marine apprenticeship to gain favour.

It was not to gain favour but to pursue the second source of knowledge that I became a mariner: To travel to the east in Venetian galleys, to the south on the Barbary coast, to the north past England to Iceland towards Thule; to fill myself with the names of sails and riggings and ports and cargos, galleon and galeasse, and so to come to know the sea and man's way on the sea such that I might begin my discourses with learned men.

Of course, like every student, I lied. But lying well is the best education. Lying early in my life has given me an opportunity to tell truths now.

*That I am not Christopher Columbus but an orphan cabin boy
taken in by the privateer Guillaume de Casenave called 'Coullon'
or 'Colombo Zovane'.*

It was he who took my name. Men of my stature are plagued by
such thievery.

That my three days alone, nineteen years old, floating
on an oar in the Ocean Sea, left me unbalanced
and full of strange resolution.

On the day that would become the first day, the sky was clear and the wind bore down east by northeast.

Our four Venetian galleys, by some called flamandes, fresh-laden from Corinth, Candia and Cadiz, bound for Flanders, worked north by oar and sail. Our dog began to bark at the bow, to the Venetians a sign of a storm, when seven flags above seven sails approached, smokepots darkening the view. The call to arms was slow, the battle long. Many of the dying prayed in French but soon they drew close by entanglement of hooks, and the corsairs threw burning oil, giving no thought to treasure. The fire was devouring, a more terrible element than its sister the sea, and quickly spread to the embracing foe. All was as Satan wished. A twenty-hour sea fight with 300 of the crew, two captains and some number of gentlemen lost. I fell into the sea and was swept away in darkness, my eyes filled with flames.

The clouds did not move. The broken oar that bore me reeked of burnt flesh. This was the second day.

I drifted through a new sea of green-dyed silk, our cargo from Baghdad, ostrich feathers from Macedon, rock alum of Constantinople, wax from Barbary melted down to crusts atop the sea in the flame of battle, Corinthian currants scratched from rockfilled fields, here sailing past my nose and there a deep red bloodstain moving ever closer in a sea that did not move. Was this the blood of those who died? The terror stopped my brain. Surrounded by the same blood I awoke. The wine-tinted sea of Homer licked my face. I was drunk with Malmsey, our cargo from Crete, flavoured with the stench of battle. It was still the second day.

The severed head of a dog floated by. The voice of Saint Christopher came from the clouds. 'I baptize thee with the waters of the ocean sea and charge you to carry the Christ beyond them.' I saw Saint Christopher Cynocephali, servant of

Dagon, Judge of Hell, tempted to seduction by Saints Nicaea and Aquilina until he converted them and they were martyred. I saw Christopher see Dagon fearful of Satan's name and so Christopher became the servant of Satan. He saw Satan fear the Cross and so went to serve the Cross. And found the hermit Pabylas by a stream who said to serve the Cross, a man must pray. I heard him answer: I know not how to pray. And so he served Pabylas ferrying pilgrims across the stream till one day a child approached and asked for passage. 'If I choose to go, I shall go. If I do not choose to go, I shall not go. Yet I choose to go.' And unknowing, Christopher bore the Christ Child and his weight, the world, across the stream.

I saw my mother carrying the child I was, from church to church in Genoa and there outside of each in black and white, these words: Christophori sancti faciem quicumque tuetur, Illa nempe die non morte mala morietur. Whoever shall behold the image of Saint Christopher, this day shall not fall or die.

I had fallen but I had not died. The sea rose in anger. It was the third day.

The cargo had all scattered and as land approached, crashed ashore. I clung to my piece of wood, now stranger than before, and not knowing if I was to land on shores known or unknown, recited the words of Isaiah: Arise, shine, for your light has come, and the glory of the Lord is upon you. And nations shall come to your light, and kings to the brightness of your rising.

I awoke in a room in the royal city of Portugal.

*That my ability to predict success was a result of my having
crossed the Ocean Sea in 1482 and found the lands then that I
would later discover as if new.*

Some of my enemies believe I made my first voyage ten years
earlier than truth. Others have heard the tale of the ship's pilot
who died in my arms in Africa after telling me of his voyage to
the West many years previous, and of the prevailing winds and
currents needed to reach the new lands there.

A few, with knowledge of the seas, think I sailed south and
east around Africa, and continued on to India and kept on sailing
east to arrive back in Spain, thus ensuring my ability to sail west
later and arrive back where I had already been. And a very few
are certain I have captured the soul of Ulysses, and simply wan-
der.

To you, reader, I will admit that all of these tales are in some
way true. I had made my first voyage a thousand times in my
imagination, on the maps of my brother, drawn the route with
sticks in the sand, imagined the winds in the sails and, in my
dreams, talked in unknown tongues to strange people. What man
of vision does not know where he is going and what he will do
when he gets there?

Some claim that I and the brothers Pinzon simply kidnapped
some Moors, put them on empty islands off the coast of Spain,
forced them to build rude huts and paint themselves like savages
and then pretended to discover them. Of the outrageous charges,
this is my favourite.

.

*That I am a very simple man, given to reaching obvious
conclusions from what I see or hear.*

I understand this charge. Those who make it accuse me of seeing
the corpse of an Oriental man washed ashore in Ireland and
immediately recognizing that I must sail west to find Cathay and
Jipangu. Or that I saw carved sticks washed ashore in the Azores
after a westerly wind or driftwood of trees that do not grow in
Christendom.

These accusations are true and not true. I have seen these and
other signs and recognized in them their truth and their useful-
ness.

Foremost, I saw that God wanted to give me the use of evidence, the evidence of my eyes, that could then be incorporated
into my arguments.

Imagine how many times and to how many people I repeated a
description of my enterprise, and presented argument, meeting
their objections and their counters with the words of scholars and
with explanations of my calculations of the circumference of the
earth. But none of this was as convincing to many of them as
when I described the bloated, water-logged body I had seen,
described the thick eyebrows, the eyes of a shape quite unknown
in Christendom but known to me as Oriental from Marco Polo's
descriptions, and from drawings in various sources.

This caught the imaginations of my listeners. They could then
start to understand such a wonder as this: that the Orient must
be so close that a man could drown in Cathay, and be buried in
Ireland.

I tried to make my arguments not so much simple as direct. I
recognized that I must fashion arguments to appeal to the knowledge and interests of the particular listener.

The Queen has a character much like mine. For her, like me,
the expedition's highest purpose was the making possible of the
End of the World. With her, I may talk as freely as with our

49

mutual friends, the monks of Saint Francis, through whose intercession the Queen granted me audience and whose belief in my mission sustained me through difficult years before Their Majesties became my sponsors. With the Queen, I may talk from my heart, and I believe she talks from hers.

The King, her husband and my sovereign, tempers his love of God and acknowledges God's part in the expedition with the practical realities of carrying out God's plans for the conquest of Jerusalem and Constantinople. He considers the investment that must be made, and the rewards that will flow into the royal treasury and the army that such funds will allow us to raise. I speak to him in the language of power and influence and history.

There are two kinds of men of commerce with whom I deal. Some, like the wise Luis de Santangel, understand the mission but still assess the returns they should expect from their investment. They are sufficiently clever to recognize the importance of my scientific arguments but sufficiently practical to perform the calculations, to assess the cost of losing three small ships stocked with provision for ninety men.

The second sort of merchant has no use for science. He wants only to be assured that I am not a madman, to know that someone else of influence believes in my story. This is the man for whom the bloated corpse and the driftwood are important because he must have some argument that he can understand enough to tell his father or his brother or his wife or his banker. This is the man who will invest little, and not in the higher vision of the expedition, but in the simple economy. This is the man who will invest only because Pinzon says so, or Santangel, or the King. This is the man who expects to earn a great deal from his paltry investment.

That I abandoned my wife and children to pursue my fantasies.

My wife, blessèd be her memory, Dona Felipe Perestrella, abandoned me early on to take her place in Heaven.

My son Diego never left his father but travelled to all the courts in Spain with me. From Cordoba to Salamanca and Granada, while what you call my fantasy turned to beggary. We played at Duke and footman starving in the halls of the Duke of Medina Sidonia. He coloured the charts I made for the Duke of Medina Celi. We played at Franciscan and Dominican while awaiting the judgement of the Royal Commission at the convent of San Esteban. We played at Moor and Christian in the soldiers' camp during the siege at Granada. And we played at King and Counsellor there as the Second Royal Commission decided that my visions were fantasies. We kicked a ball together and threw sticks at birds, hoping for a meal, on the muddy road to France. He saw the true game of Beggar and Friar played at dusk before the monastery at La Rabida until Father Juan Perez ordered the doorkeeper to let the shoeless boy rest the night and, for the boy's sake, the father too.

This night reversed my fate, for the boy explained my plans to the good fathers while I slept, exhausted and defeated. Thus say the Genoese: 'Make the son heir to your thoughts, fortune will follow.'

Other, lesser men have abandoned their fantasies to pursue a woman. Others, cleverer than I, say that fantasy is a woman. I have had the pleasure of meeting several of these women.

That my voyages were religious expeditions disguised as voyages of business. That my voyages were business trips under the guise of sacred, holy expeditions.

These are not charges made by intelligent men.

I have shown elsewhere that there would have been no religious expedition had God not provided me with the language of investment potential with which I could approach His Majesty and those like him who had access to wealth.

There is a comedy in this which has not escaped me. Men provided wealth and hoped for greater wealth in an expedition whose goal was an end of wealth.

That my search for a Royal Patron was simply the attempt of a greedy man to rise in status and that I never intended to sail in search of new lands.

I have discussed this accusation and those similar with my brother Bartolomeo, who reminds me of our debt-ridden, sorrowful days when all the envious saw were two brothers visiting the courts of Europe flailing colourful charts and telling tales of preposterous voyages to gain favour and treasure and a better name for a low family of Genoa. He has urged me not to respond since now, what the envious feared has come true, but through our deeds, not words.

Yet I shall tell the story of my first encounter with the Queen of Spain.

After the Commission of Cardinal Mendoza had investigated my theories and claims and plans for a second time and found them worthy of consideration, I was taken by Luis de Santangel, Minister of Royal Revenues, to the camp outside Granada where the armies of Castile and Aragon were besieging the Moors. He took me to the Royal tent, reminding me of protocol and that I would see the Queen alone as the King was inspecting the tunnels under the walls of the city.

'Señor Colombo, you do not strike me as an Admiral.' I knelt and bent my head to kiss the bottom of her dress. 'It is customary to kiss the sword of the Queen when her house is at war.' I did so, still thinking of how an Admiral should respond. I rose to speak.

'An Admiral does not respond until asked. Now, if you are trustworthy and your theories so correct as my advisors tell me, why have you no treasure to fulfil your plans?'

'Do you imagine I am not an honest man because I have no money?'

'I imagine no man honest who has plans until he has completed those plans. Until then, there is much benefit to being dishonest.'

53

I had no response. I suspected even my religious arguments would bear no weight with a Queen in armour.

'Besides, the King has no money. Yet his plans are very grand.' She laughed again. 'Come Admiral, sit and speak. Your plans are more interesting than most and the benefits you ask are no burden.'

That I alone was responsible for the delays in having my first voyage approved because of my reluctance to argue with the learned judges of the Royal Court who were convened at Salamanca and at other places.

In the year of our Lord 1486, Her Majesty the Queen Isabella established a tribunal of scholarly men to advise Their Majesties as to the value of my proposed enterprise of exploration. I might have said that these worthies were to advise on the practicality or feasibility of the voyage, and to some extent that was the case.

They were also to advise on the scientific foundation of my project, and this was a matter of consternation to me. Their questions were indeed frequently of a scientific nature, the critical one: 'Since God created the world several thousands of years ago, and many great men have lived since then, all contributing to our store of knowledge, how is it possible that they would have never yet discovered what you now set out to discover?'

I spoke, but could not express the true nature of my intents. I recognized those moments where I had revealed too much of my true purpose, and recognized that the scientific answers that they wanted to find were simply those topics that were the least interesting.

They know less about the wind than I, less about currents, about sailors and about those things known and unknown in the Ocean Sea. Their science could be described in summary as 'How is it you could know that which we do not?'

Father Gorricio reminds me that I continue to sound somewhat bitter. I recognize that too.

That I spent ten years squabbling over the negotiations for my voyages instead of simply voyaging like a true discoverer.

Anyone who makes such an accusation has never attempted to mount an expedition or to be an entrepreneur. Just as great men throughout history depend on financial backing to forward projects which some believe dubious in nature, so did I depend on the fiscal pledges of my sovereigns.

It is irresponsible to voyage without funds for supplies and without proper authority. How can one conquer new lands without authority? In whose name does one sail? By whose authority will one govern?

By my establishing a partnership with Their Royal Majesties, by enjoying their influence and good will, I ensured the success of my future voyages and their place in history. And, by coincidence, ensured food in my and others' mouths. In Genoa, it is said 'Men's ships sail on their stomachs.'

*That I do not show sufficient respect for the wise and noble
council charged with determining if I should be granted
the Queen's authority to sail.*

In Salamanca, Father Georges de Metez, an abbot of Valencia,
was walking through the market and met there by chance one of
the learned judges.

The judge was a short man, but he carried a carved walking
stick that was nearly as tall as his shoulder.

My friend the priest, who is not given to rudeness, com-
mented on the exquisiteness of the carving of the cane's ivory
handle.

'Yes, it is pretty,' said the judge, 'although it's a great disappointment to me that this cane is too tall.'

'But you can cut it to the right length,' said the holy father.

'But the handle is the most valuable part of this cane,' replied
the judge.

'Of course I agree with you,' said the priest, slightly puzzled
by the response. 'You should cut it at the other end.'

'Father, I've thought of that,' said the wise man chosen by the
royal court to decide my fate, 'but the bottom of the cane is not
the problem. It is too tall at the end with the handle.'

Father Metez was sufficiently surprised that he felt compelled to
invent a great pain in his knee so that he could quickly take his
leave. He found me in my quarters and related this tale.

That I have been blessed with the power to perform miracles,
indeed that I counted on miracles in times of great difficulty.

Would that this were true. God has accomplished much through the agency of this, his humble servant, but not the rendering of miraculous works.

I believe we live no longer in times of miracles. I believe that the learning of our wisest men makes it no longer necessary for God's presence to be felt in and through miracles.

If the facts of my story were different, if when first I came to Her Holy Majesty the Queen in 1486, Saint Catherine had spoken to her and immediately my request for assistance had been granted, that would have been a miracle.

But I would not have recognized the miraculousness of such a decision. Indeed, the Queen's later decision to support me despite the objections of the Learneds, seems now to me to have been miraculous.

I know that it was not a miracle.

I mean no blasphemy when I say that the events and decisions that led to my first voyage were of greater significance than if they had been miraculous. This seems extraordinary, I know.

I expected men to recognize the Divine Nature of my intentions and simply to say Yes, it must be so. Instead I was asked questions to do with the width of a degree, a question whose answer would imply the circumference of God's round earth.

The Queen's learned advisors were attempting to determine whether I should sail in pursuit of God's Will by assessing the likelihood of correctness of my answers to questions about the circumference of the earth. Let me repeat myself in order that you, my distinguished readers and auditors, will fully appreciate the momentousness of their analysis:

The Queen's learned advisors were attempting to determine whether I should sail in pursuit of God's Will by assessing the likelihood of correctness of my answers to questions about the circumference of the earth.

Neither I nor Their Majesties recognized the terrible implications of the Queen's request. The Queen, a woman as free of sin as any human may be, asked her noble scientists to pass judgement on the merit of My Holy Mission, as if their cold calculations could somehow assess the Holiness of my Cause.

Even Pilate was clever enough to recognize that he could not pass judgement on Our Lord. These learneds, in their vanity, asked me questions about the nature of winds. Perhaps, in their hearts, they realized that they had no place passing judgement on me and my mission. Cowards that they were, they told Isabella that my voyage must not be supported.

Holy woman that she is, she offered to make me Admiral of the Ocean Sea. But I fear for a world in which small men of fact are asked to place value on the actions of those of vision.

*That my relationship with the Queen of Spain was based
on more than my theories.*

Without theory, men and women would have no relationships.

One day in a Royal Library scattered with my charts, maps and tables, the Queen said 'You are an interesting man, Admiral-to-be, Admiral-yet-to-sail, Admiral he knows not where.' She rose, surrounded by her skirts, from the floor, where I had laid the charts, attempting to explain to her the tendencies of the winds. She threw some of my papers into the fire. She was easily bored and the room was large and cold.

'You are a man driven to go, where most men only stay. In the beginning they stay to prove their love. Once love is given, they stay to earn it still. Later, they stay only to consume love. They stay because the surroundings are pleasant. They stay because it is time for a meal. They stay for the children. They stay for the Church. They wake up and instead of going, stay to sleep. In the end they stay because they are cowards. Even those who go to war, stay. And I have heard from friends that those with mistresses stay forever. The very fact of having a mistress proves the cowardice, the fear of going. You, it seems, have no such fear.'

'I have no mistress, Majesty.'

'Not yet, Admiral-to-be, but neither have you gone.'

Accusations Concerning My First Voyage

That I was ill-prepared for my voyage of 1492.

No other explorer in history has been as well prepared as I. At a very early age I entered upon the sea navigating, and have done so continuously until now. The calling in itself inclines whoever follows it to desire to know the secrets of this world.

I have traversed every region which is currently navigated. I have conversed with learned men, ecclesiastical and secular, Latins and Greeks, Jews and Moors, and with many others of other sects. The Lord made me very skilful in seamanship, He gave me sufficient knowledge of astrology and of geometry and arithmetic, and He gave me an ingenious mind.

The Lord has set me to live in those regions which would be most useful to my education, to live amongst cartographers in Lisboa, to learn about the winds of the Ocean Sea in Porto Santo. I have studied all writings, cosmography, histories, chronicles and philosophy, and through these writings Our Lord gave me to understand that it was practicable to navigate from Christendom to the Indies.

There is nothing that God or I could have done to prepare me better.

That I deliberately fostered the myth that I was sailing off the edge of the world when I knew the world was a sphere.

I have seen the shadow of the earth, round, move across the face of the moon during its eclipse. I know the world is round.

So too does the rest of the world, and has since the days of the ancients. I know the origin of this story of three small ships sailing off the edge of the ocean.

The servant girl Susana Fontanarossa told a certain old man in her village of her dream in which she sailed west across the Ocean Sea with me. (She did not know my name at that time, or my mission, she knew only she would sail with someone who looked as I do.) This old man told her that she would be heading into a world of leviathans and two-headed men, of Atlantis and Antillia. He warned her that at the edge of maps there were terrors, real and imagined, terrors far beyond the imagination of the cartographers who created them. He said that she must be brave to go first.

In this private conversation between us, I did not hesitate to tell her I wanted to sail to this edge. This world created by mythologizers and mapmakers. The edge of the maps.

That the true credit for my expedition should go
to Martin Alonzo Pinzon.

The family Pinzon, long before I met them, had sufficient wealth to mount their own voyage to the West. Certainly they would not have done so without the authority of Their Majesties, but surely a wealthy family, with its own ships, sailors, and provisions, could have petitioned, and would have received such authority.

If indeed they possessed 'secret knowledge' of routes to the Indies, as some have claimed, if indeed they understood the circularity of the winds, if indeed they could have performed this expedition, why then did they not?

I give them credit for their abilities and their wealth and their support amongst the sailors of Palos de la Frontera. Without them, my departure might have been delayed by several months, and for this I am grateful.

I cannot, in all honesty and candour, describe them as men of courage.

That I was presumptuous and without science.

There are learned men who cannot see with my eyes, who cannot accept that such an unschooled sailor as I can have a plan for a voyage of great import. They ask questions drawn especially to reveal mistakes or errors in my plan.

I answer in the words of Saint Matthew, 'O Lord, who wouldst keep secret so many things from the wise, and revealest them to the innocent.' And Saint Matthew says that when our Lord entered Jerusalem, the children sang 'Hosannah, Son of David'. The scribes, in order to try Him, asked Him if He heard what they said, and He replied, 'Yes, dost thou not know that from the mouths of babes and innocents, the truth is pronounced?'

I am just such an innocent.

That I am a very complex man, full of complicated motivations.

I have never understood this charge.

Some say I was motivated by gold. This is true. One serves God by finding sources of His most Precious mineral. I recognized that Their Majesties would be unable to mount a Holy Crusade without new sources of wealth.

Fifty years from now, if this book still finds readers, they will perhaps not understand how dangerously important it was in my time to recapture Jerusalem. You who read this book in that future time, when Constantinople and Jerusalem are once again glittering jewels in the crown of Christendom, you will have forgotten the urgency felt by all the faithful to bring about this triumph. I sailed West so that our Christian armies could march East.

You cannot imagine my pain when Constantinople fell to the Turk, and my joy when Granada embraced Christianity. I wanted to experience that same outpouring of success in my heart again and again.

I thought then that I was no warrior. I knew that my greatest contribution would be through gold. Through achieving success in a direct trading route to Cathay and the lands of spices, Holy Spain would be able to raise the most powerful army on earth.

I know now that I was born to lead that Holy Army, but this is new knowledge, and was not part of the motivation for my first voyage.

*That I invented the meteor that fell before my ships
on September 5th, 1492.*

Fire in the heavens has always been a sign from God. The men were given signs of hope frequently enough that they were able to tolerate my vision.

During the evening of September 5, I was reminded of the prediction made by Susana Fontanarossa on the dock at Palos on the morning of my departure.

'There are only three types of expeditions,' she said. Although she knew that I would not let her sail with us, her mood and her attitude seemed not to change. She acted as someone would who knew that she would be proved right later.

'And two of those are the same,' she continued. 'When merchants sail, the holds of their ships are bursting with the goods they wish to exchange. The expedition is self-contained. It doesn't matter to anyone where they go, or what they bring back. They are preceded by arrangements, by legal agreements, by payments and promises to pay.

'When armies sail, the situation is no different. The holds of their ships brim full of the weapons of war and they bring back spoils that could have come from anywhere, that, under other circumstances, would have been the result of successful trading. Such expeditions are preceded by proclamations, by declarations, by wailing and gnashing of teeth.

'The third type is that of the explorer. His holds are bare, carrying only the provisions needed by his fleet. He transports no trading goods; how could he? He knows not whither he sails and what the market at his destination will demand.

'You think that what makes your little fleet special is that you have no maps to guide you, that you are three tiny specks alone on a great Ocean Sea.

'No, what makes you special is that your holds are empty and that you do not know your destination. Your expedition can be preceded only by comets and showers of meteors.'

That I ordered Juan Arias, who had chosen to disobey my express orders and bring with him aboard the Niña a small cat, to throw the cat overboard on our leaving the Canarias as a sign to the men that I would tolerate no disorder in my fleet.

No. This is not true.

That I lied to my men, keeping two sets of records of the distance we had travelled, telling them only the shorter distances.

Instruments of measurement at sea are inexact, relying on crude observation of the sun and the horizon. Mariners who are skilled recognize this inexactitude, and maintain records that show both minimum and maximum numbers. There are men who believe that instruments of scientific calculation cannot exaggerate; such men keep one set of records.

It is true that I publicly announced, at the end of each day, only one number, and that it was the smaller number. My men were not of such sophistication that I could have told them two numbers each day of our distance travelled. They were crude men, for whom it was necessary that their captain be recognized as having the authority of his instruments. These were men who believed in these instruments. All the other captains with whom they had sailed had always given them to understand that such numbers were exact, and accurate, and reliable. I could not, on one short trip, dissuade them from this point of view.

I understand their desire for such figures. They had spent days sailing in a direction which no one yet had ever sailed, and there are no landmarks on the sea. They (and I too, of course) could only track the distance and time we had behind us. Unlike every traveller after us, or ourselves, we had no maps to indicate that the next morning we would reach our destination, or three days hence, or even a month.

So each day I performed my observations and did the calculations.

Knowing their temperaments and the sea, I thought it wiser to tell them the lower numbers. These numbers were as likely to be true as the higher ones.

That I planned no entertainments for the sailors nor thought to bring women on a voyage that was of indeterminate length.

It is a code of the sea that women on a ship lead to more unknown than a pilot without maps. I, of course, as a Captain-General of that voyage could have changed the code.

And sailors, reader, are always entertained. We played for hours at *tarrochi* and other card games, *scopa* and *sette bello*, from the Genoese shipyards. As well, I did hear rumours that there was devised by them a special entertainment on this voyage, called by the vulgar, Sleep with the Queen.

It consisted of a coin and a fish cut in half.

A gold maravedi, a coin of the realm with Queen Isabella's visage on one side and the King's on the other, was affixed to the mainmast each day before the evening watch. The sailors who were then free from that watch took turns throwing their knives to see who would be closest to the coin. After much raillery and rude laughter the winner was agreed upon, usually decided by my brother sighting foolishly through his quadrant from his table in the aftcastle. Then the chosen sailor was given the largest fresh-caught fish saved from that day's meal. The sailor took his knife from the mast, kissed the coin, sliced the fish cleanly in half and proceeded to sexually ease himself, either in the mouth of the fish, held tightly in his hands, or in the red fleshy centre of the back half of the fish, a hole sometimes being cut for that pur-pose, depending on his interest that evening.

Sailors of course are naïve from years of hardship and sailing and do not realize that there are more ways than two to please a woman.

I heard that such events took place, but witnessed none of them and took no part in this particular entertainment.

That I stole the pension due to Rodrigo de Triana, who, some claim, first sighted the New Lands.

I have never stolen any man's wealth, title or due. If there had been good reason to give Rodrigo de Triana credit for the small discovery of an island in the night, I would have. But in reality, from the forecastle deck of the *Santa Maria*, I saw a faint light at the distant horizon toward the end of my evening watch. Unsure of my discovery, reluctant to raise an unnecessary alarm with a crew tired of my constant messages of imminent landfall, I told the night watch to seek further evidence.

Rodrigo de Triana, on the *Pinta*, four hours later, acting on my orders, saw the land whose lights I had seen. The sailor Triana has since become despondent, travelled South or East and become a follower of Mahomet.

The pension, income from the Royal Slaughterhouses in Sevilla, was not significant.

That my arrival in the new found lands was without ceremony,
and unbefitting.

On October 12, year of Our Lord 1492, I stepped out of a small
armed boat carrying the Royal Banner and placed it on an island of
green trees, much water and many fruits. Beside me were the
brothers Pinzon, carrying the flag of the Admiral, the Verde Cruz,
a green cross signifying the unity of Christendom and Islam in
Jerusalem, and beside the cross the letters F and Y, the benefactors
of this unity, Ferdinand and Isabella. I ordered Rodrigo de
Escobedo, Notary of All the Fleet, to bear witness and testify that I
took possession of said isle for the King and Queen.

Then a great many people of the island gathered. They were
dark and naked and afraid of our beards. That they might feel
great friendship for us and because I knew they were a people
who would better be freed and converted to our Holy Faith by
love than by force, I gave them some red caps and glass beads and
many other items with which they were greatly pleased. They
called their land Guanahani and Jivao, and pointed in all directions. This led me to believe, at that time, that Jipangu was
within reach. I resolved then to take six of them from here to
Their Highnesses that they might learn to speak Spanish and easily
become Christians as it appeared to me they had no cult of their
own. That they were not unwilling travellers, despite the accusations of my enemies, and they are many, you shall see.

I spoke these words as I planted the Royal Banner: 'O Lord,
Eternal and Almighty God, by Thy sacred work Thou hast
created the heavens, the earth, and the sea. Blessèd and glorified
be Thy name and praised be Thy Majesty who hath deigned to
use Thy humble servant to make Thy sacred name known and
proclaimed in this Other Part of the World.'

I recognize that they spoke no Spanish, but the Indians seemed
grateful that I had just claimed them and their land for Holy Spain
and much appreciative of the ceremonial aspects of the moment.

That I had no authority to assign names to creatures, plants and places that I discovered.

My advanced years and the strain of my legal battles are affecting my powers, once strong, of recollection.

I remember having this argument with Susana Fontanarossa on my first voyage, on our first landing, at San Salvador, which the natives called Guanahani, or Ouatiling.

And yet I know there were no women on that voyage. She could not have been present.

I had seen a kind of thick-skinned tuber and plucked it from one exposed root and announced that we would call this food 'cassava'.

She laughed, and shook her head at me, vigorously, and when she had stopped laughing, said, 'If you are Adam, then I shall be Eve. I also want to name things!'

I was silent with surprise.

'In what book did you read,' she continued, 'that you have authority to thus fabricate names? What if this plant already has a name of its own, and always has had? What if it grows in Palestine, or Sicily, and has had a name for many hundreds of years, a name which thousands of people have spoken before you and of which they are very fond?'

She took the tuber from my hand, and placed it next to her ear, and motioned to me to be quiet. 'You are very lucky,' she said finally. 'It is a cassava.'

That I know what I wish to hear from the natives when I question
them, and lead them into answers that agree with my previous
understanding.

Far easier to discuss geography with a man who has read Juan
Mandeville than one who has not. Far easier to learn how to find
the objects of one's explorations from someone who speaks
Spanish, Italian, Latin or even French, English, Portuguese,
Gaelic, Hebrew, Aramaic or Arabic, than someone who speaks
only the language of the Indies.

I have taken our ship's barge to many beaches where I have
stood face to face with stern men who spoke no language I or my
translators zecognized. I have stood on the ship's deck looking
towards eighty-foot-long war canoes made of a single log seeking
direction and instruction from stern men who had never read
Juan Mandeville's account of his travels nor any other book, and
who displayed a surprising lack of interest in the affairs of the
Great Khan who ruled some several hundred leagues to their
West. Nor did they evince interest in or knowledge of commerce, or mining, or gold.

Yet despite this, I had no choice, serving God and Their
Majesties, but to press for the answers I needed.

On one beach in early November, thinking myself only 100
leagues distant from the cities of Zayto or Quisay described by
M. Polo, I pointed to the south and one of the Indians howled.
With gestures I described the men with dogs' heads that Juan
Mandeville said live on islands off the southwest coast of Asia. By
various signs, the Indians indicated that such men lived on an
island seven days distant by canoe.

Not content to learn only so much, I described in greater
detail the record left by Juan Mandeville that such men eat the
flesh of other men, drinking their blood and cutting off the genital parts for trophies of war.

The Indians, by signs, gave me to understand that such was

75

the case, and that they much feared these creatures. I gave them beads and other trinkets to indicate that we were their friends, and would afford them protection against such cannibals.

'Cannibal' or 'carribal' is their word for the flesh-eaters.

At this point in our discourse, Luis de Torres, a translator, gave the Indians to understand that we sought also men with one eye, also described by Juan Mandeville. They indicated that such men were far away, pointing to the south.

This encounter continued over many hours. Certain old men told us that in a place called Bohio there was an infinite amount of gold, which was worn on the neck, in the ears, on the arms, and on the legs; and also pearls. I further understood them to say that there were great ships and merchandise, all this to the south-east.

Although the Indians told us these things, I felt that they were paying attention to these subjects only because of our questions. There was no indication that they themselves were involved in commerce. I asked if they were at war with the Great Khan, whom they call Cavila, and they indicated they were.

Years later, in the garden of the monastery where I write, I cannot help but marvel at what we learned on our first voyage. We never found one-eyed men, or men with dogs' heads, or a great harbour with hundreds of ships.

But when the Indians told us there was gold to the southeast, we sailed to the southeast. We found parrots.

A man would tell us of a great palace to the northwest with many rooms in it, and when we sailed there, we found a village with cotton nets slung between trees for sleeping, called by the natives 'hamacas'.

Luis de Torres made gestures of a bird's wings to one group of natives, and told me that on the next island there were men who could fly. Then he rowed back to the *Niña*, and brought a cannonball to the beach, held it in front of his face, gestured, was gestured at in turn, and announced that there were men on the second island distant who had no faces, but only cannonballs for heads.

Juan Mandeville wrote of many places, and of many strange peoples. Much of what we know today of geography comes from his books.

I wonder now, years later, whether Juan Mandeville saw the one-eyed men himself, or men with dogs' heads, or merely described them in gestures to people with whom he had no common language, and to whom he gave beads and combs and mirrors when they said yes to any of his questions.

*That I had no expectation of meeting the Great Khan; I would
not carry trinkets of metal and glass to offer such a ruler.*

The trinkets and bells and brass and pots and combs, scissors,
needles, hats, mirrors and coloured cloths were a blessing, and a
source of much good. They were such that every ship carries,
whether voyaging the one day from Cadiz to Alicante, the four
from Venice to Crete, or the many down the coast of Africa.
These are not truly trinkets but the sailor's and his ship's every-
day needs from the times when sailors first sailed from Phoenicia
to the Pillars of Hercules. How else would one toll the watch,
prepare the noonday meal, repair the sail and signal the battle
formation? What else to provide the more personal entertainments to which sailors have ever put each sundry object found
aboard their vessel?

Susana Fontanarossa, in her wisdom, has told me that I never
expected to meet the Khan, that two centuries have gone by
since the time of M. Polo, and the Khan is dead, or a man who
exists only in stories, like the African Presbyter Johannes.

She tells me, and I allow her to tell me these things, and I do
not become angered, that I have always known I was discovering
a New World, but that the pain of such a momentous discovery
was too much for me, and that I deny it in order that I can better
be at ease with my failure to predict the existence of such a
world.

It is easier, she says, to understand that one is sailing towards
the known, than the unknown, and that my real strength is that I
sailed towards the unknown without realizing.

She reminds me that my arguments to the Learned Judges at
Salamanca and elsewhere were based on my determination that
the circumference of the earth was as predicted by Esdras and
Aristotle, and that if the lands I discovered are not those of Asia,
then my calculation of a degree is wrong.

Several points must be true. The Indies lie between Europe

and Asia. We do not know at what distance. Clearly, I have discovered new lands, the outer regions of the empire of the Khan, between our own and Cathay, which lies just beyond. My brother Bartolomeo has drawn the isthmus of Veragua on his map at the eastern edge of the continent of Asia and I agree with him. At the same time, it is equally true that I have discovered so many hundreds of islands that it may be said that they constitute a New World, one which appears to pay no tribute to the Khan or any other ruler.

The Indians have told me of the distant cities and the gold mines and of the palace of the Great Khan. I have seen none of these sites.

Nonetheless, I am a man who stands by his degrees. Susana Fontanarossa admits (and tells me that I recognized this intuitively, in my silence) that it is far easier to find financial support for a voyage to the known by a strange route than to the unknown with any strategy.

That my voyages forced the pernicious division of the world by
the imaginary Line of Demarcation between Portugal and Spain
through the Bulls of His Holiness Alexander VI
on 3rd and 4th May, 1493.

The world has always been divided. By the Greeks among their gods. By the Persians into satrapies. By the Romans between their Caesars.

In your villages and towns, you have drawn a line around your neighbours' land. The Pope, Rodrigo Borja, a Spaniard, has simply drawn a line around the world. A line drawn, some say, to appease the jealousy of the King of Portugal who, upon seeing in my discoveries the truth of my projections, regretted having dismissed them. The Pope's line is yet another melancholy example of what may happen when the Church co-operates with the spirit of the times. All will stand and fall in cycle until there is one world again united under God.

Until those times, the world and its peoples remain divided, whether by imaginary lines or by religion or wealth or position or age.

The chief who slaughtered the settlers I left behind at La Navidad was no less wise than many of the wise men of Christendom. Gesturing in a sweep to all the islands and his island, he said, 'You think only that this is the land of our ancestors. We think it is the land of our grandchildren.'

*That the Pope has no authority in the lands I discovered, that his
jurisdiction extends only to those creatures descended from Adam
and Eve, and those plants and animals descended from those
of the Garden of Eden.*

'How,' asks a man after Mass, a man who thinks of himself as a
religious, a man whose name I have forgotten, 'how can these
new lands be considered as part of this creation? I have seen what
you have imported, and heard what you have described. There
are flowers never before seen in Christendom, and trees, and animals, and, yet more puzzling, men and women unlike any others.'

'They are most unlike,' I reply, 'at once more savage and in
certain ways, more refined.'

He continues: 'The Holy Testaments, Evangels and Epistles
tell us of one God, and of one Holy Family in the Garden. Further, the Holy Word of Moses tells us of the flood of Noah, and
how all the animals were saved.'

'I have spent time among these creatures and these people,' I
say. 'I recognize in them God's work. I recognize too that He
has led me to them in order that they might be saved.'

'Impossible. This is some strange part of the world abandoned by God and under no obligation to our Pope and our
Rulers. You blaspheme even by trying to incorporate this Indies
into Christendom.'

'You have not seen it. I recognize God's Holy Work in the
flowers, in the trees, the animals, and in the people. Naturally,
then, His Holiness has dominion.'

'Nonsense. The natives of the Indies are descended from
another Adam. You have no right to have any commerce with
them, under any circumstances.'

I resisted a great temptation to comment on this man's own
ancestry. I recognize that there are many such as he, and that
they cannot be reasoned with.

*That I was prepared to sacrifice all that I owned,
all those whom I held dear, and all in which I believed, in order
to ensure my place in history.*

Is there a role for an individual in history? Does one rise to the challenges of a moment in the broad sweep of events and bring one's person (one's strengths, one's weakness, one's peculiar, particular, personal history, and one's ability to recognize and follow God's guidance and will) into a confluence with all the circumstances of other lives being lived at that moment, and thereby achieve a greatness of purpose?

Had I been born five centuries later or earlier, might I be or have been a peasant or a wool-comber, and no Admiral at all, and contribute only to the rush of events in my quiet way, and not be given by God a task of the magnitude He chose to give me by virtue of His recognition of the applicability of my talents to the challenge of these my times?

Or does one simply and thankfully play one's part knowing that God's purpose will be fulfilled, and if not by one agent, then by another?

How did I change history? Did I change history? There were others, certainly, who, because of their reading or their experience, reached the conclusions that I did regarding the distance from Spain to the Indies.

But if their experience had not taken them to the equatorial zones (as mine had), how would they know that constant good winds blow in that zone, both to the west, to achieve the journey, and to the east (further north) to return?

Although the expense was great, there were others, certainly, who could lease three ships, or who owned them and could outfit them for such a trip. They did not.

Many people talked of this route. Maps existed which showed the islands of the Indies just where I found them.

But what combination of vision and Divine Guidance and pigheadedness and fortune thrust upon my unworthy soul the task of accomplishing these works?

What credit do I deserve for the fulfilment of God's word as prophesied by so many Saints?

God and the Saints found me righteous. They found me knowledgeable in the ways of His seas and man's ships. They gave me the strength to persevere against all charges and accusations against me.

They continue to do so.

Yes, I was prepared to sacrifice much, no less than Abraham, had I been asked. But I never sacrificed my belief, and my place in history is as nothing when set against my place in Heaven.

Of course, I cannot deny my role in history.

When I first presented my enterprise to Their Highnesses, the King and Queen, history changed. In their wisdom, and because of their interest in the pursuit of knowledge, they granted me audience and thereby established the validity of such an enterprise for all the rulers of all our neighbouring lands.

Many mocked me, and my enterprise. Many wondered that the King and Queen should devote any attention at all to such a cause. But at the very least, the Western route to the Indies was now open to discourse amongst individuals beyond the circles of mapmakers and sailors. Now, for the first time ever, merchants and men of commerce could argue the merits of such a route and such an enterprise.

This was accomplished merely by the fact of Their Majesties' interest, long before they decided to do me the honour of sponsoring the enterprise.

When I sailed on my first voyage to the Indies, history changed again.

For all purposes of practicality and governance, I sailed for the Empire of Spain. But many other nations sent envoys to watch my departure secretly, to try to ascertain from my provisions how long a journey I intended. And they waited to hear of my failure or success, knowing that whether or not I returned, whether or not I found Jipangu, they would all benefit from the knowledge accumulated on this voyage.

New maps would be drawn. Books would be written. Some hundred sailors would now have the experience of sailing west across the Ocean Sea to those lands that had previously always been to the Orient.

Whether they knew this fact or did not, I sailed for all of them, with all their eyes on my back. What I accomplished, thanks be to Our Lord, I accomplished for all of Christendom, and indeed for all of the world, for I enabled our Christian Fathers and Brothers to go forward into all the world, as Jesus commanded that they should do, that all men might be saved.

Thus could the Indians hear the Gospels. I cannot take the credit for this, sinner that I am. I only opened one door. Now all the Christian lands will flock together to bring the Peace of Jesus to the Indian peoples.

My place in history? To humbly open one small door.

How did I change history? I enabled that door to open in the Year of Our Lord 1492, instead of two years later, or twenty years later. I opened that door on the Indians of San Salvador, instead of Indians to the South or North, or instead of sailing a distance further to the West to come upon the people of Jipangu or Cathay. I opened that door for Spain, and for His Holiness, the Pope.

And I opened that door then, in that place, for Jesus, Mary and Our Lord, and for all their Saints and Apostolic Missionaries.

Accusations Concerning the Enthusiasm of Christendom for My Second Voyage and the Events of That Voyage

That on my return I used my fame to mislead Queen Isabella.

My triumphant return to Granada was actually a small parade of
some of the local guilds, a few Royal functionaries and three of
the Queen's liveried footmen, after which I was led into the
library of Her Residencia. I saw the disappearing crimson cape of
Cardinal Mendoza, by some called the Third King of Spain after
the Queen and her husband, sliding out of one of the secret
doors in the room. The Queen was dressed in green and had her
back to me as if to go. I could see she was holding a small note.

'I understand, Admiral, that you have discovered three islands
and saved six souls.'

She turned and sat as I approached and knelt to kiss her skirts.
'I have discovered a new world, Highness, and there live more
new souls than God can count.'

I placed on the floor two presents I had brought back for her
as evidence of my thanks: a golden disc from around the neck of
the most beautiful of the native girls and two feathers from the
cucu bird, captured and named by Susana Fontanarossa. The
Queen seemed pleased for a moment.

'You have delayed in Tagus and stopped in Lisboa to see the
King of Portugal.' She was not pleased.

'Tagus was my port of return and I stayed only to tell the tale
of my voyage to the Prince of Navigators, as is the custom of the
friendship of mariners. And may God forgive me for creating
envy, but I went to see that King to remind him of his mistake in
rejecting me.'

'And are you here to remind me of my mistake, Admiral?'

'Your Majesty has made none.'

'None that you know of. But be aware that I know of yours.'
She walked towards the fire and threw in the note. ' These gifts
remind me of a boy I knew in my youth, a boy who pestered me
with love and gifts. The gifts were childish, whatever he could
find on his way to me. Flowers from a field, bits of leather, broken sword hilts, old potatoes, small rocks. The balcony of my

father's castle was littered with these things. Soon my heart was charmed and my father having gone to war, we made love each night on the balcony. You must tell me some day if that is like making love on a ship's deck. This boy would be gone each morning when I awoke, leaving some new trinket on the balcony. But one day he was still there in the morning. He proclaimed this true love and said he was going into the village to find me a true gift and would be back. I dreaded his return. By nightfall he had not come and I have never seen him since.'

'Is this to tell me about gifts, or love, or mistakes?'

'Are you staying, Admiral, now that you have returned?'

I spread my charts and tables on the floor and took out the new maps made by Bartolomeo and together the Queen and I planned the second voyage.

*That, further, I misled Their Majesties and indeed all those who
sailed with me in the second, third and fourth voyages,
as to the riches they could amass in the new lands,
thereby enticing speculators.*

I am not responsible for the extraordinary jubilation that greeted
my return from the first voyage. Spain had been at war and had
defeated the Moors at Granada. The Jews had been expelled. All
the most ambitious sentiments of glorious Christian empire were
ablaze.

'There are twice as many people on these ships as live in the village which is my home,' said Susana Fontanarossa, on the morning of our departure on my second voyage. She sailed as a settler
when I refused to let her wear the clothes of an ordinary sailor
and to act as one of the men.

'No one would tolerate life in such a village,' she continued.
'More than three-fourths of the people will do no work, and the
rest will take care of accomplishing all that must be done.

'I admit that when we arrive at our destination, they will have
work to do, to stir together the ingredients we have on these
ships with the ingredients we find there, and to build a colony, to
build a town that makes them feel they are still in Castile but that
looks in every plank and every face as if some part of the ingredients are fresh and new.

'In the meantime, on this ship, they are like cargo. There is no
historical precedent for how useless they are. In some ways, they
are like pilgrims en route to the Holy Land, except they do not
pray enough. They have the same festive glee, the same shared
nature of their purpose, but, unlike pilgrims, these ones expect
that everything will be given them. They are not like scholars,
who travel to visit the great libraries and learn from conversation. They are not like gypsy tinkers; they have nothing to sell.

'They expect to be shown beautiful islands, glistening with
sustenance and shelter and wealth. They expect to be given a har-

bour with a townsite, there to build a miniature Granada or Sevilla.

'This kind of people has never before existed, so I have invented a new word, in their honour, to call them. I have named them ''passajeros'', passengers.'

*That the crews of my ships were maltreated and that I allowed
them to suffer unduly in the pursuit of God's Will.*

Ten days out from Spain we were becalmed for two days and
two nights, when a small boat bumped up against us in the dusk.
Father Perez and I were by the wheel. He had said the evening
Mass for the men and women aboard, and as was our custom
after each day's mass, we were arguing gently about the prophecies and the signs that were being fulfilled by our voyage.

We heard a slight commotion, and presently two men from
the *Estrella* were brought to us, one the mate, whom I recognized, and the other a settler, a tinsmith by trade.

I had barely nodded to the mate when the tinsmith, fervently,
passionately, took my hand and kissed my ring.

'Señor Don Noë,' he said, 'Blessèd are you, and blessèd am I
to be blessed with you.'

I looked at the mate for an explanation. 'He has decided you
are the ancient Noah brought back to life,' the mate said. 'For the
most part, he wanders around the ship smiling and congratulating
himself, but several times today he has burst into loud lament for
his family and his home village of Tardona. He is upsetting to the
other passengers and as you know, our little caravel is very
crowded. Our captain thinks too that some of the other passengers are starting to take seriously this idea, and have begun crying
quietly for their families, who they say are all drowned.

Before I could speak, Father Perez turned to the tinsmith.
'What can you tell me of the flood?' he asked.

'Father, I understand it all,' the man said. (I have forgotten his
name.) 'Here, in the middle of the sea, surrounded by water, we
are in the only place where it will not rain in times of flood. Our
home lands are being flooded, but we do not even see rain clouds
on the horizon behind us.'

His voice quieted, nearly to a whisper. 'I began to realize, a
few days ago, that this Ocean was rising higher and higher. Soon I
realized that as it rises, it will flood the coast behind us, and as it

continues, soon all of Iberia will be flooded. I became frightened last night when I looked into the sea and saw the tops of mountains below us.'

He turned to me.

'Please, Señor Don Noë,' he said, 'I have done some humble calculations. I have calculated that by now the waters will be almost to the height of our village in the mountains. We have left our little son there with his grandmother. Please order the captain of our ship to turn back so that we can rescue him. By my calculations, there is just enough time to sail directly to the mountains. He cannot swim.'

Father Perez asked the tinsmith quietly: 'How do you know that this man is Noë?'

The man answered quickly, 'He told us before we sailed that God spoke to him. Then yesterday I went down to the animal deck and saw two horses, and two goats, and a sow and a swine, and a ram and a ewe. Then I understood.'

'Did you sleep last night?'

'No, Father, I watched the tops of the mountains beneath us. I could see also trees, and roads, and villages.'

'And the night before, did you sleep then?'

'No, Father, I don't think so. I was calculating how much rain must fall to flood our village.'

'Let us go and read together the story of Noah and his ship,' said Father Perez to the man gently. 'I will make you a cup of tea, and in the morning we will visit the animals below.'

He turned to the mate. 'Please come back tomorrow,' he said quietly. 'The man will be well by then. Our crew will make certain that we have an odd number of animals on our ship by morning.'

*That I ordered Juan Arias to throw overboard his lute and
Rodrigo Minio his hurdy-gurdy in order to demonstrate that I
would tolerate no ungodly performance of music on the voyage.*

No. This is not true.

Seventeen ships! Filled with men and some women, all of
them bristling with the anticipation of adventure and wealth. The
mood was that of a festival, except that it was spread across
seventeen ships, sometimes separated by several miles, usually by
several shiplengths.

That is: The ships were close enough that shouting and merriment carried across calm waters, and induced similar behaviour
in nearby ships. In fact, Juan Arias and his lute were on my ship,
and Rodrigo Minio travelled with his hurdy-gurdy on the *Ber-
muda*, another man had a tabor on a third ship. On a calm sea,
they performed music together, accompanying each other across
the small channel that separated our ships.

On this trip, amidst the festivals and singing, the young red-
haired woman, Susana, with my mother's name, Fontanarossa,
sailed with us. It was she who said that our ships were not from
Spain but from Venice.

This seemed a peculiar statement, and when I asked her to
explain it, she admitted she had never been to Venice.

'We are a city,' she said, 'which floats on the sea as Venice or
Amsterdam does. Unlike people living in other cities, we never
leave our homes, which are large buildings like palaces, and
which contain all our servants who cook, and our animals, and all
our friends and enemies.

'But we know what all our neighbours are thinking. We know
when they are happy, because we are always happy at the same
time, and when we are bored, they are bored. We wave at each
other from time to time, and we know they think just as we do
because they wave back.

'Sometimes we wonder whether there is really just one ship.
The other sixteen are perhaps simply reflections, as if they exist

only in mirrors. We are surprised to discover this since every ship looks different, until we remember that when we look at ourselves in a looking glass, and slowly turn our heads, we see many different reflections of ourselves.

'Sometimes, on clear, calm days, we can see that there are thirty-two other ships, and when we wave at someone on one ship, people on two wave back. Sometimes we realize that half the people waving back are upside-down, sailing under the water, but sailing at exactly the same speed as those of us who sail through air.

'This is unusual,' she concluded, 'because the ships which are sailing upside-down have no keels.'

That I took no books to the New World.

I have been accused of believing that there is no need for libraries in the Indies.

I reply that there are two arguments: Which of the books of our old world matter to the Indies, save the teachings of Christ? Do they care about Alexander the Great? Julius Caesar? The emperor Constantine? They have their own stories and histories, relayed to their children without books. What of the Great Khan? When we meet him, shall we tell him about Portugal and Holland, or only of our own glorious Empire? Does he want to know that there are wool-weavers in Christendom? Tinsmiths? Diseases? Lutes? Mountains? Lemons? What matters to him whose empire is full of extraordinary sights and goods and who lives quite happily in ignorance of us today?

At the same time, I recognize that one cannot build churches without libraries. The priests whom I brought to the islands are missionaries and teachers. They believe, as I do, that the path to conversion is through knowledge, that the human creature thirsts for religious understanding as it does for water, and that the more one learns of the ways of Christendom, the more likely one is to follow those ways.

It starts with learning to speak Spanish.

That I proposed the capture and enslavement of the natives of the Indies.

This was a desperate and despicable proposal and the Queen, in her goodness, had the wisdom to dismiss it immediately. The speed with which she did so shocked me into realizing my folly.

I made the suggestion that we enslave the cannibals who were terrorizing the gentle and generous Indians whom I counted as friends. I wrote the letter quickly, in a fit of desperation and energy, in the frenzied belief that perhaps trade in human souls was the only business I might establish in the lands I had discovered. Foolishly, I ordered slaves sent to Spain without waiting for a reply from Their Majesties.

This was a temptation placed in my path by the powers of darkness, and I succumbed. For those few months, I imagined that a successful trade in slaves would prove the value of the Indies, and make suddenly worthwhile all my years of preparation and suffering.

I did not then know that my suffering was just beginning.

When I arrived in Barcelona immediately after my first voyage, I was greeted by Their Majesties as Admiral and Viceroy. Thousands of children, women and men lined the streets I walked up to the Court. No one cursed me, no one called me a foreigner or a dreamer. They recognized in me the Great Discoverer that I am.

Within three days, the Pope himself knew of my discoveries and had divided the world between our Holy Spain and the Portuguese and it was I who had claimed the Spanish half for our Majesties, and who had claimed all the new lands, for God and the Pope.

The next months saw all of Christendom set ablaze with the fire of missionary zeal, and, it must be said, with greed. Imagine that two years earlier the court's learneds had dismissed my plans for an expedition, that one year earlier I had been faced, albeit briefly, with the prospect of filling my ships with condemned criminals. And now, great bankers were offering me money to find a place on my seventeen ships for their eldest sons!

Seventeen ships! God's ways are mysterious. We provisioned and found crews for seventeen ships in a matter of months.

But beneath my euphoria, I was gripped by a tiny knot of fear.

I have a tendency toward enthusiasm, and it is true that my reports to Their Majesties stressed the docility of the Indians and their amenability to Our Faith, and that I stressed the availability of gold, convinced that I had seen such small amounts only because I hadn't come closer to the gold mines themselves. The forty men I left behind at La Navidad would surely have discovered the source.

My fear came from my recognition that no one heard me when I spoke. My sailors discovered that the more they exaggerated their tales of gold and the beauty of the Indian women, the more envious were their friends. Men who had not been on the voyage told me of mountains made of gold and silver, which reflected the sun at noonday so that one's eyes would be burned if one looked, and one had to turn one's head away for fear of losing one's sight. A Holy Friar asked me if it was true that one had only to hold the Holy Bible before the chief of an Indian tribe, and immediately all his men would kneel and pray and speak in tongues.

I didn't know what to tell the Holy Father. I assured him that the Indians spoke in tongues and turned quickly down the next passage.

During my first voyage, I worried that none of the men shared my dreams of our eventual success. During the second voyage, I feared that each man had his own dream, and that none were alike, and that no land, new or old, could survive comparison with these dreams.

I had no idea how deep would be their disappointment, and how dangerous would be the resentment of men who felt I had promised them gifts of land and gold, and now, when they arrived, was refusing them.

So many of the wrong men came to the islands.

The ones who came to work – the farmers, the weavers, the carpenters, and so on – worked and were satisfied. They lived as well, perhaps, as they had lived in the towns and villages they left behind them.

Those who came to pick gold up off the streets were disappointed. There were no streets.

Perhaps at that time I was one of those men. Perhaps I believed the stories I had heard about the wealth of the Indies.

I know only two things. That when I saw the rubble of the fortress of La Navidad, and not one corpse, and heard how the men had been slaughtered, I was changed in that instant. The voice of Satan was in my heart, cold, and he told me this was a cruel land, that there was no gold, that the Indians could never be Christianized, and that the seventeen ships anchored offshore, and the 1500 people, and the horses, and the swine, and the cattle and the goats, and all we had brought with us here, all this had been transported for no reason. That somehow I had tricked them, or made a mistake, and that there was a wind that had brought us here which was just air and breath, and it was my breath and my voice, and that if I stopped talking or breathing, they would all glide gently back to Cadiz, and the cold in my heart would melt.

Later we found some bodies, and some parts of bodies.

The other thing I knew was this. That anchored offshore were 1500 men in seventeen ships and they were expecting me to act. Many thought I would order the slaughter of the Indians who were still on the island. I did not. Indeed, at the time, the idea didn't occur to me as a possibility. Many thought I would rebuild the fortress, only stronger, with better weaponry, with more men. I wanted only to leave that place, and we sailed, immediately, to the east.

Further east on the island of Española, Martin Pinzon had reported that he had found much gold. At first I thought to calm the men by finding that place, and showing them thereby that they were right to follow me to the Indies.

But one cannot explore a coastline in a fleet of seventeen ships. We came, after only a few days, to a good harbour, and good land, and began building the town which I named Isabella, to honour Her Majesty and to bestow God's blessing on this place.

I thought often of the Queen, and wondered what I could tell her in the letter I had begun to write. I could say that there was no gold mine, that some Indians are more violent than those I had seen on my previous voyage, that our men had been slaughtered and La Navidad destroyed, and that our colonists were disheartened. I could say that we hadn't yet found the Great Khan or his ambassadors, had seen no spices of any kind, nor even any of the great cities of the empire about which M. Polo had written. I could say that all we had seen were islands, and that we had been unable simply to turn a corner and find Cathay or Jipangu, or any continent at all.

I chose to say some of this and not say all.

With the voice of Satan in my ear, I succumbed to his temptation and proposed that there was one good commercial reason why we were in the Indies, and that was to create a trade in native labourers. Indeed, as is my custom, I acted quickly, and ordered that many natives be captured, and they were, and they were sent to market in Spain, where they were sold and many died.

The King and Queen ordered me to be arrested, and I returned to Spain in chains, God's punishment for my fall into temptation. I wore those chains in penitence.

Later, as you know, I renounced this proposal and denounced this position. As proof that this statement is true I offer the evidence of my continuing friendship with the Franciscan Fathers, good and holy men who denounced the enslavement of Indians then and later.

To my great and everlasting joy, sinner though I am, never again did I fall victim to the terrible temptations of the powers of darknesses, and I have walked all my days in God's path.

*That I put this plan into action, capturing and enslaving
the natives of the Indies.*

Though several years have passed since I wrote that letter, I
remember still the care with which I attempted to explain my
plan.

I was concerned by the inability of our colonists to attract the
sustained attention of the Indians in order that we might teach
them Christian ways. My strategy was a simple one: We would
remove some of these Indians to the Catholic soil of Spain where
they would be immersed in the Spanish language and in Christian
teachings. I suggested that we would undertake this operation
most particularly with the warlike Carib Indians, thereby also ridding the friendlier Indians, for whom I had the utmost fondness
and respect, and ourselves, of our most dangerous adversary on
the islands.

I recognized that Their Majesties would tire of provisioning
the new colonies with men and sustenance, and it seemed to me
that the exchange of the Carib for our livestock requirements
would fulfil many goals at once.

The conversion of disbelievers into Christian men and
women, however it comes about, is always of crucial importance.
My proposal, which seemed injurious at first to the condition of
the Indians, would surely have been seen as appropriate and necessary when the first of the converted slaves would return to the
islands as Christian teachers, ministering to their fellows, and
fulfilling the prophecies.

That I and my fleet brought only pestilence and disease to the New World.

We brought oranges. And lemons. We brought bergamots and melons, chickens, calves, goats, hogs and sheep. We brought horses.

We brought grain seeds, seeds of herbs and of vegetables, grape vines, sugar cane, and a multitude of grafting trees and saplings.

We brought medicines and salves and lotions and powders and pills.

We brought small hawks' bells and mirrors, beads of coloured glass on threads, needles, pins, combs, copper ornaments and brass wind chimes; knives, coloured caps, bowls and cups, metal eyelets for lacing shirts and shoes; all these items to be given as gifts to the Indians.

We brought gold and pearls, small silver jewellery, cinnamon and pepper and other spices, all these items to be shown to the Indians that they might understand that these were the gifts we sought in return.

We brought carpenters, joiners, stonemasons, roof-makers, miners, men skilled in all trades necessary to create settlements in a New World, men skilled in administering, planting, cultivating and harvesting.

We brought works of art, paintings and sculptures.

We brought books. And paper and quills.

We brought maps and map-makers. We brought sails and masts and skills in navigation.

We brought a man expert in accounts, and a lawyer.

We brought the Word of God.

On our second voyage, we brought back five of the six Indians who had accompanied us to Spain on our first voyage. They had been baptized with great ceremony in Barcelona, sponsored by none other than Their Majesties and Prince Juan. (The sixth stayed behind in the Prince's household, at the Prince's request, and died shortly after, the first of his people to be welcomed into Heaven.) No doubt, they in turn brought good news of their treatment by our Sovereigns to their countrymen.

We brought, from the Queen's own chapel, the magnificent gold crucifix, her chalice, and such ornamentation and vestments as would befit a church outfitted in her name, reflecting her great desire that the gentle Indians be ministered to by men of faith.

We brought, from his Holy Father Alexander, twelve zealous priests, led by Father Buyl, appointed by the Pope as his apostolic vicar to the New World.

We brought crossbows. And lances, muskets, arquebuses, gunpowder, and swords, cannons and munitions.

We brought clothes. Pantaloons, shirts, chemises, undergarments, hats, helmets, stockings, shoes, boots.

We brought persons of all ranks, and of several countries. English, Portuguese, Dutch and Italian joined us. On our second voyage, amidst the great enthusiasm for the expedition of all the people of the Holy Nation of Spain, we sailed with not only the 1000 persons hired, but with 500 volunteers and stowaways clamouring to be allowed to take part.

On seventeen ships and several small boats too, the three largest each carrying 100 tons of burden. To build a New World. To use the tools and the bodies provided by God and Their Majesties to build a New World.

That I slept inconsistently and at the wrong times of day.
That I refused to sleep for many days at a time and then slept
almost incessantly during those times that my leadership was most
profoundly needed.

What a strange and purple charge is this.

It is absurd that I should be accused of such things and yet more extraordinary that such charges should have been related to my friends the good priests and to the King.

I was offered sleeping draughts on several occasions by a certain sailor who fancied himself first a chemist and later a barber, but I always declined them. The man intended me no harm. He became a settler in Española and enjoyed a moderate success as a barber, administering an extraordinary and very colourful collection of herbs and powders of various combinations and efficacy to all who sought him out. He married the daughter of the Indian cacique and healer whose techniques he imitated, the old man Palahuelo who was my friend, and whom I sometimes called Kee.

That I mocked religious observance by wearing sackcloth in the streets of Sevilla upon my return.

Perhaps I am the first man to combine the virtue of humility with the necessity of arrogance.

After my second voyage, filled with gratitude for the mysterious ways of Providence, and recognizing in myself the sinner who had been blessed through God's choosing, I found it appropriate to don sackcloth with a group of Franciscan penitents, my brothers all, and to walk in procession through the streets of Sevilla.

I had become used to men recognizing me when I go walking in public, and shouting 'Dreamer', or 'Admiral of Vanities', or 'Admiral of Mosquitoes', or 'Foreign Climber'.

But some men in the street mocked me with peculiar taunts, and, to my shame, I cursed them. I recognized immediately that my heart had hardened after so many years of impatience at beginning the expeditions and now so many years of constant work in keeping them alive.

What was it that took me so by surprise? Their doubt of my religious sincerity: I heard someone shout 'False Penitent' and then 'Lying Breast-Beater'.

*That I exaggerated the results of my second voyage to maintain
the favour of the Queen and that these exaggerations continue to
this day.*

'I have discovered eleven thousand islands, Majesty, and named
them for Saint Ursula and the Eleven Thousand Virgin Martyrs of
Cologne.' I showed her Bartolomeo's maps of the voyage indicat-
ing the Virgin Islands.

'There are not that many islands here, Admiral. But I find no
fault with this. There were certainly never eleven thousand vir-
gins in Cologne. Nor in any other city in Christendom. Neither is
a virgin necessarily a martyr.'

'I rarely exaggerate, Majesty. Perhaps this is a good occasion
to begin.'

Accusations Concerning the Trials and Difficulties of My Third Voyage

That I left to Englishmen the possibility of discovering lands that should have been Spain's.

'There are four cardinal directions,' said Susana Fontanarossa, as we sailed from Sanlúcar de Barrameda on the first morning of my third voyage, 'and we are very lucky. When we reach a certain point in the Ocean Sea, perhaps as soon as three days hence, we will be able to sail in any three of the directions and be exploring new territory again.

'The Portuguese who sail south from their homes have the choice only of south or west. On the east they are blocked by Africa. Only the English have the same choice as we.'

I interrupted her.

'In principle, you are right, but I follow the advice of Aristotle who said that those things in the world that are best, and most valuable, are those which come from the lands where the sun is hottest. We sail west and south.

'Those who sail from England sail west and north, if only to avoid us. They cannot be considered Aristotelian.'

That God, in the Divine Persons of Mary and Joseph and Our
Lord, made express the divinity of the relationships which make
up a family, and that I had no right or authority to transport
fathers and sons of families to New Lands to become colonists
thousands of leagues from their families.

Those who so accuse me live under the delusion of an old picture of families. Look to the number of your sons and brothers and fathers who are sailors, or merchants, or knights, men such as I am, who sacrifice the joys of home to accomplish those tasks to which God has set them.

Those who accuse me of making colonists out of their sons and fathers sometimes go on to accuse me of inventing the idea of colonies.

This is equally foolish. There have always been Gitanes, and Jews, and Moors who travel from the land of their birth to settle elsewhere. Indeed, there are many souls of Spanish origin who live today in Jerusalem, or in other parts of Christendom.

I and my family, all Genoese, have found a place in Holy Spain, to which we are devoted and loyal. We, and others, will always seek out new lands in which to settle where new opportunities arise. Our name in Italian means 'dove', the creature that returns to Noah, having found land.

It is coincidence only that my family name, in Spanish, is 'Colón'.

*That I gambled with the lives and souls of sailors by leading
them on a voyage that could have been their last.*

No sailor ever thinks any voyage will be his last. No more than a
dying man thinks a certain breath will be his last, or a child imagines a sunny day will be followed by a storm.

Yet thus did my wife, who suspected the future, both hers and
mine. Her last words were 'Where are you going?'

Indeed, that those who choose to explore the world instead of their souls offend God.

I believe myself to be a rational man who fully espouses the scientific nature of our times. I welcome the advances that are being made with the invention and dissemination of the printed book, and in the areas of navigation, of astrology, of cartography, of physics, of alchemy, and the related sciences.

Strangers and friends both are frequently astonished by my enthusiasm for these advances, feeling that one who seeks the truth about the workings of the world pushes beyond what God has chosen to reveal and acts at odds with the will of God.

Let us assume for a moment that such criticisms are reasonable and have a sound theological basis. How, practically, do we live in such an era?

What God revealed to Adam was a Garden, and His Will, and a world beyond the Garden. What He revealed to Moses was His Land, His Word and His Law. To others he has revealed other, smaller parts of His Plan.

Far from being criticized for exploring beyond the edges of what God has previously chosen to show us, I would have been far more remiss had I refused, had I announced that what God has thus far revealed is all that He intends us to know.

Imagine Moses standing at the foot of a mountain in the desert saying: 'No, I will not climb. Surely all that God intends us to know, we know already.'

꧁꧂꧁꧂꧁꧂꧁꧂꧁꧂꧁꧂꧁꧂꧁꧂꧁꧂꧁꧂

That I tolerated stomach-talking amongst my sailors, and made no effort to prevent its spread.

Curiously, the outbreak of this harmless but annoying pastime in my fleet began, innocently at first, with a dove that perched on the rail above the wheel before we left harbour.

I had never seen such a thing, although Susana Fontanarossa, the Queen's former servant girl, said that doves commonly make such tricks. He (or she, perhaps) was sitting quietly making a low, warbling sound. Suddenly the bird looked to the railing at starboard, and it seemed that the sound was coming from there.

We were all very surprised, because there was no second bird at that place. The dove turned quickly to the other side, and now the sound seemed to be coming from that direction.

Susana Fontanarossa laughed and said that this was stomach-talking, that it was an ancient art, and that humans could perform it as well as doves and other birds.

'Most horrible!' she cried suddenly, and ran toward the starboard rail. From the water below, we could hear someone calling for help, a female voice.

We could see no one, but the cries, muffled, continued. One of the men was preparing to jump into the water, when I noticed Susana Fontanarossa was laughing still.

'Stop, sailor, do not jump!' I shouted to the man. All the crowd now realized the jest and Susana Fontanarossa became a teacher of stomach-talking to many of the sailors and settlers. She claimed that there was no mischief in this activity, but that anyone on a ship for several months needed diverting.

Luckily, their enthusiasm for this foolishness lasted only a few weeks, but during that time, it was very difficult indeed to know who was speaking and to whom. Once or twice, others, attempting to speak in my voice, gave orders that should not have been obeyed.

That I failed to visit my father upon his death bed in 1498.

I did not know of my father's death until 1499. I was elsewhere, about my Father's business.

OOOOOOOOOOOOOOOOOOOOOOOO

That I worship the sun and the stars, putting them before God in my heart, and performing complex rituals in their honour.

I know the source of this picture and am myself to blame. At noon one day, six days south of the port of Isabella, just before discovering Trinidad, the Indian Palahuelo (also called Kee) observed me taking readings with my sextant and fell to his knees.

I did not interrupt my task to invite him to stand up as I should have, but allowed him to continue to scrutinize my gestures. I completed this ritual daily and frequently he took part, kneeling, silent, immobile, and neither curious about my actions nor able to distinguish them from the rituals we performed each Sabbath Day.

Kee sailed later to Castile in the company of Father Perez and lived for some time with the Franciscans, whom he found 'full of surprises'. He was gracious with me always, and I consider him a great and good friend.

That I lived my life in fulfilment of prophecies and therefore was always looking backwards.

'Many voices have said that our voyages open the door to the future,' said Susana Fontanarossa, sitting, curled on her own legs in the large armchair which my men sometimes carried to the aftcastle deck for me on clear, crisp nights. There was no moon. Illuminated by the lantern Bartolomeo had suspended above his drawing table, she looked eighteen years old again, with the vigour and forthrightness and green brightness of eyes that I remember still from the first day we met, a decade earlier, at Palos. We were at anchor off Paria, Bartolomeo adding notations to his new map, I resting against the wheel.

'But yet so much of what we do,' she continued, 'is our attempt to resurrect what is past, to achieve an earlier state in history. For many centuries, people have sought the gold mines of King Solomon. We have found them. We have found the Garden of Eden. And we seek still the way to take the mission of Christendom to all peoples in the world. Our goal is to undo the effect of the episode of the building of the Babel Tower and recreate that state in which all men and women are baptized to the one Catholic faith, the allegorical equivalent of the time before Babel when all spoke in one language.'

I felt compelled to respond to her backward-looking description of my mission.

'The march of history,' I said, feeling that the healthy exhaustion of a good day's exploration was freeing up my own reserves of allegory, 'is like the cauldron of fishes and shellfishes that we often enjoy for dinner.

'In the beginning is the fish or the langoustine, pristine, whole, holy in its beauty. This state is akin to the days of the Holy Parents of us all in the Garden of Eden.

'Then the chopping begins. The food is still uncooked, spread across the block, fish and onion and cassava and other vegetables all mixed in together. They are tossed in the seasoned water, and the second phase of history begins, the boiling, and the slow simmering.

'This entire analogy exists only to make one small point: The morsels with which we began are the same ones which we continue to cook. The Garden of Eden, the mines of King Solomon, prophecies foretelling the importance of islands and that all the world should be saved, and so on – certain big pieces bubble back to the surface from time to time, more succulent and more flavourful for having cooked longer.

'One adds new ingredients and new flavours over time – we ourselves have contributed – and eventually all is cooked, the great fish stew of history, ready to be served to Our Lord at the Final Banquet.

'But you mustn't be misled. Some of the ingredients are ones you recognize from the past, but the stew continues to simmer here in the present.'

There was silence for a few moments. Constellations both named and without names sparkled in a broad sweep from the deep blue dome above us.

Finally Bartolomeo spoke.

'I'm still hungry too,' he said. 'I often feel hungry a few hours after dinner here in these islands. Perhaps some of the cakes are left.'

That I built gallows at Vega Real to hang rebellious Indians.

I built crosses, 340 of them.

That I tolerated disorder in the town I founded.

I believe this accusation is based on the unusual actions of Guillermo Bacca, a settler, a window-maker by profession.

A man specializing in the particular glass needed to make fine windows, in a site where a town is being created from sand and forest, such a man expects to be very busy, and to earn much.

Dismaying then is the realization that the perfect weather of the island affords all those who build houses the opportunity to have simple openings in their walls, perhaps with shutters, certainly without glass.

Guillermo Bacca brought his clay pots and a supply of lead, he built a furnace, gathered the cleanest sands, potash, and materials for colouring, and waited for business.

Eventually he constructed a beautiful window for himself, and then built the frame of a one-room house near the mouth of the bay, with no walls. He installed the window, and sat behind the frame, looking to the east, towards his home in Madrid. He sat like this for much of each day, for many months, until eventually I offered him passage home.

'No, thank you,' he replied courteously, 'I prefer to watch from here.'

That I betrayed my Royal Benefactors and their representative Bobadilla.

'Well, Admiral, it seems we have a problem.'

My brother and I were in manacles in the stinking hold of *La Gorda*, ready to sail for Spain as prisoners of the Crown. I swore to him then that these chains would forever remain at the door of my home. They are there still.

'Forgive me, Admiral, but preserving these manacles will not get us out of them.'

'Commendador Bobadilla has done this as representative of the Queen. Therefore I shall speak to the Queen. Did you see the signs posted in San Domingo depicting me as Admiral of Mosquito Land?'

'I saw worse, Admiral. I saw drawings of you as Admiral of the Queen's Bed. But these Castilians are proud and lazy. Two traits that are often found together. They expected the Indians to do their work. Your orders for them to work alongside the natives offended them. They see you as a tyrant, and tyrants belong where we are.'

'You were Adelantado, you could have forced them to obey.'

'I was busy converting the natives. Did you know some of them were clever enough to be baptized many times? Each occasion was a feast day, a day on which they did not have to work.'

'This founding of new worlds is tiresome business, Bartolomeo. The natives do not want to be Christians and neither do the Christians.'

'This is a lesson, Admiral. There are many to be learned here, thank God. I, for example, have learned that being of the family of a great man one inherits not greatness but enemies.'

'This seems to be true.' The ship creaked and lurched throwing me into my brother's lap. I rested there. 'Let us at least enjoy this passage. For once we are in charge of nothing.'

That after my return in manacles I was forced to regain the favour of the Queen through witchcraft, potions, and personal sorcery.

It was somewhat more than two months after our arrival when we were at last allowed to see the Queen. That time was passed in the monastery at Las Cuevas near Sevilla, here where I write today, my brother and I revising charts, planning what we began to call the High Voyage and debating with the monks the wisdom of God and the Queen. Since I had suffered imprisonment and had been stripped of my honours, the monks were not permitted to call me Admiral and so addressed me as *El Passajero*, The Passenger.

When the day came, Bartolomeo and I were taken to the Alhambra. We saw that none of our previous supporters, monks, ministers, cardinals or courtiers were there to greet us. Only Señora de Rossa, mother of Susana, also employed by the Queen, awaited in an antechamber crowded with workmen on scaffolding painting over Arabic murals with scenes from the lives of the Saints.

'Come, Bartolomeo, we shall go to the chapel and say a prayer to San Thaddeus, patron of just but unlikely causes, while your brother visits with the Queen.'

I prepared myself as I entered the library. I had no potions or magic, just one gift. The Queen was by a large window, standing by an open book on a bookstand. She was dressed in yellow and red, the colours of the houses of Aragon and Castile.

'Welcome, my former Admiral-to-be. Your voyages become varied and troubled, like those of Signor Polo, of whose travels I am reading. You understand, I hope, that a Queen cannot go, but must stay where she is, trapped by duty. I travel through the tales of others.'

I understood this well, and laid my gift, a single white pearl, at her feet. 'This comes from the Mouth of the Dragon, Majesty, a river so wild and wide that it signals a new continent rising to a peak where lies the Garden of Eden. The gem itself is round like our world and white as a pure soul.'

'I have a garden out my window, filled with flowers round and white.' She looked out the window. 'I can imagine them to signal more than your pearl.' The Queen then turned and scooped the jewel off the floor. She put it in her mouth and swallowed.

'I eat your new world.'

I pleaded with the Queen for one more voyage. I told her tales of boys from the Indies with their genital members sewn to their stomach so they could not meddle with their elders' wives, or so they would grow fat to be eaten by their cannibal captors. Isabella responded with what she thought was a third and more likely reason for this aberration: they would grow up to be priests. I was shocked by this and so told her of their temples built in the jungles of the islands to the idol, Seyti, where the high priest, not sewn, would wait each morning in his white cotton cowl. The women of the villages, who did all the work, would go with offerings to the temple. The first to go in would enter with no gift but would lie in intercourse with the priest. Then the other women would come and kiss the first and leave their offerings. The Queen mused over this, then kissed me.

'I give you one last voyage, Admiral. You may go because you cannot stay.'

Accusations Concerning the Period of My Fourth Voyage

That I have been of two faces with respect to the New Lands,
sometimes saying they are of Asia (and therefore Lands Known)
and sometimes insisting they are hitherto unknown and can be
therefore of different consequence to Their Majesties' wealth
and influence.

Certain facts cannot be controverted and they survive all scrutiny.

Men who were wise told me that there could be no new continents between Asia and Europe, that it was impossible that land of such significance could remain undiscovered so late in the history of the world. This seems to be a reasonable argument whose only objection is that God reveals all things in His time.

Readers, you who live in Cathay or Jipangu, you will be better able to judge this charge than I.

꧁꧂꧁꧂꧁꧂꧁꧂꧁꧂꧁꧂꧁꧂꧁꧂꧁꧂꧁꧂

That, out of revenge, I destroyed Bobadilla and Roldan, and their ships and treasure, by leading them into a tempest outside the harbour of San Domingo.

'Shall we tell them?'

My brother and I were standing in the water up to our necks to avoid the mosquitoes. We were not swimming, but were on a shoal where we had anchored, protected, outside the harbour. Bobadilla and his superintendent, Roldan had refused us entry to the harbour on grounds of protocol: They were about to sail out, on their way back to Cadiz.

We had noticed that morning a large storm from the north proceeding directly towards the harbour mouth. It was this storm Bartolomeo was asking about. We had opportunity to warn these people by sending a small boat.

Just then I slapped my hands together, killing a mosquito about to land on my nose. This made a loud noise, interpreted by the crew of the small boat as an order to go. These messengers implored my enemies not to leave the safety of the harbour; they refused, and sailed into the open sea. Thus perished Bobadilla, Roldan, fourteen ships and one mosquito.

That I attempted to steal from Roberto Cimicata credit for the invention of swimming.

On the contrary, my role was very small, and I admit this readily. I served simply to point out, as any knowledgeable person would have, that conditions are much different between the Holy Empire of Spain, and her colonies amongst the islands. I indicated that the most compelling argument against *natación* – I did invent this word for Roberto Cimicata's invention, as he knows no Latin – was the quality of the water, and therefore, while he could happily indulge himself on an island, he must not disport himself in the water that flows through the largest cities of Aragon and Castile.

Our ships, *La Capitano* and the *Santiago*, were pierced, worm-eaten, leaking dangerously as we headed back to Española from Veragua and Belen, from encounters with tempests, from fierce, unsuccessful battles with natives, and from losses of ships and ships' barges. The ships were in such a state that it was clear we could not reach Española in safety.

To save us all, I determined to run the ships aground on a beach in Yemyaica, and to send Diego Mendez along the coast to the point where this island was closest to Española. From there they would cross in a small canoe, and, traversing Española, would reach the town of Isabella, where he would obtain a ship in which to return to fetch us and our belongings. Since the new governor of Española is my enemy, I recognised that Mendez might have difficulty in finding such a ship.

His delays in being able first to cross the strait, and then in getting help, meant that Roberto Cimicata had nearly a year in which to perfect his invention.

It is my belief that all men and women recognize the pleasurable sensation of immersing their bodies in water. Normally, in bathing, one is standing, and the sensation is slight.

Roberto Cimicata was a fastidious bather, and, even on the ship, sometimes bathed once a week. Shortly after we were shipwrecked on the beach, he began bathing every day, with the excuse that there was little else to do. Certainly no one disagreed.

And then, on the morning of July 29, 1503, barely a month after we had been grounded, he announced that it was foolish to carry water from the sea up to the edge of the forest and that henceforward he would bathe without a bucket, in the sea itself. All the men found this mildly amusing, but paid no attention, Roberto Cimicata already known to us all as a hothead, and a man of occasional bursts of temper.

But to our surprise, instead of simply washing himself, he sat down in the water, and then lay down, with only his head above the water. There was no wind, and the water was smooth as fine wood, with only a slight occasional ripple.

He was smiling broadly. All of us gathered around him, a strange sight of nearly two hundred men standing in a half circle on the land, with the centre of the circle being a head lying, seemingly by itself, in the water.

I cautioned him against disease, but he argued that water by itself did not contain poisons, that some water carried the poisons of other substances, and that other water was clean. This water, he announced, was clean. It was evident to me that his lying in water was not a spontaneous act, but that he had been contemplating it for some time.

I cautioned him further, reminding him that nowhere in the Holy Testaments of the Prophets and the Saints is it written that people should lay themselves in water.

He replied that although he was an ignorant man, and not as well-read as I, he could recall no express dictum against it.

I thought for a moment, and then recalled that the Holy Saints have written that Our Lord Jesus approached their fishing boat at night by walking on the water.

'Surely,' I said, 'it must be taken as a sign of some importance that Our Blessèd Lord walked on the water, and not through it, and that He taught Saint Peter to do the same.'

'Walked through the water?' asked Roberto Cimicata, with a wild expression in his eyes, and then again, 'Walked through the water?' He was almost shouting.

Suddenly he stood up and began walking quickly, away from us, directly into the sea. The *Santiago*, listing slightly to her port, her masts bare, was aground to our left. We were in a wide bay mouth, and a line of palm trees rustled slightly in the distance ahead of Roberto Cimicata, and slightly to his right.

On the beach there was consternation. Many people were shouting at him to come back, to not be stupid, to be careful, to beware of strange creatures. Two men started to run in after him, but they were clothed, and turned back almost immediately, their pants and boots wet.

Roberto Cimicata walked away from us until the water was just above his waist, and then turned and walked parallel to the beach, towards the *Santiago*.

He was obviously trying to walk quickly, and he looked somewhat amusing, slowed down by the water as if he were being held back by unseen ropes.

At one point he did an abrupt turn, and now walked the other way, sometimes propelling himself with his hands, which he cupped against the force of the water. He was smiling, and saying 'Yes, yes, yes,' more to himself than to us.

He made another volte-face, and walked back towards the ship. Two of the other men had taken off their clothes: he noticed this and shouted to give them courage.

Soon there were six men water-walking, the last of them saying to me, 'It is very hot today. Perhaps this will cool us down.'

Emboldened by their camaraderie and shouting, Roberto Cimicata walked into deeper water. Suddenly there was a small splash, and his chest and neck and head were gone. I feared he had been attacked by a monster of some kind, but he reappeared, flailing with his hands and kicking with his feet.

It seems that the surface of the earth fell away rapidly at that point, and he found himself with nothing but water beneath his feet. This was the moment when he invented swimming.

*That I did not deny being a god when the natives of the new
lands acclaimed me one.*

In the mid-afternoon, a group of fishermen have returned from
the sea, bearing their fresh catch. Their wives and children clus-
ter about them on the beach. The day is like any other October
day.

These people understand that they are at the western edge of a
large sea, and they recognize, or have heard, that there are many
other islands and larger lands, to their west and to the south.
They have met some of the people who live on these islands, but
have themselves never travelled further than the nearest island.

They have neither heard of nor imagined lands to their east.
On this day, they see the tips of the cloth sails of three small cara-
vels, any one of these little ships larger than any other structure
they have ever seen, and coming from the east.

Eventually men shining with the reflection of sun on metal
land on the beach, with much pomp and flourish and speaking in
strange tongues.

This scene was repeated many times, on many islands,
although soon it was clear that the news of our presence amidst
these lands had preceded us to several islands.

Why should it be a surprise that the man who commanded
such vessels and such a troop of well-outfitted men, why should
it be a surprise that Indians, seeing us for the first time should
take us for gods?

I was as surprised as they. I expected a centre of trade in the
Khan's empire, not an outpost that seemed to owe him no allegiance. Here I stood, and my men also, astonished and somewhat
dismayed at the small group of people admiring us.

They recognized a miracle, not I. They recognized too that
the miracle was of my God and not theirs.

I needed time to think. I needed time to understand where I
was, and to establish how these people fitted into my view of the
Khan's empire. It was only later that I recognized that God

moves through me, and that these people would be our first conversions to the faith of Christ.

Three voyages later I stood in dark of night on the bow of my ship, grounded for many months in the Bay of Santa Gloria, with candles all around. The Indians who had been offering us supplies and sustenance were beginning to hesitate. Our food was becoming scarce. It was clear we lacked authority with them.

I prayed for the moon to disappear. The minutes passed. I had managed to gather the natives, and my sailors together, with the explanation that I would create a miracle.

I prayed for the moon to disappear. To perhaps be more truthful, I howled for the moon to disappear, as a wolf or a dog might. I put on a show of calling God's attention to this little corner of his earth, to this little band of shipwrecked sailors and sceptical Indians.

Time passed. I continued to howl and to pray as if I knew exactly how long I had to do so. In fact, I did know how long to weave this spell: I had to keep praying until the thing occurred.

I saw God's hand pass over the moon on February 29th, 1503. Like the natives of the Bay of Santa Gloria, I thought the eclipse a miracle. At that moment, my God became theirs. The fact that I needed an eclipse to quell restless souls, that I had the *Almanac of Regiomontanus* on board to predict the eclipse, that all was compounded by being the night of the day of the Leap Year, makes it no less a miracle.

The Indians were thus encouraged to continue to provide us with food. We waited more than one year for our rescue to occur.

Perhaps a more interesting miracle would have been to rescue us earlier.

That human nature argues against the truth of my story of the eclipse.

Certain people to whom I have related this episode doubt that such a small act of drama could have been responsible for such a change in the Indians' attitude to us, reminding me that eclipses appear with regularity, and that the Indians must surely have seen previous ones.

This is certainly true, but after my demonstration the Indians believed that the eclipses they had seen at other times must also have been warnings from the God of the Christians, which they had failed to understand.

We can see from this story what it means to be men of science, as we are: the Indians did not believe it was possible to predict on earth what would take place in the heavens. We recognize that our God is not diminished by our ability to predict some of His actions. Indeed, perhaps this is the true miracle.

*That I allowed my men to continue to swim when I knew
it was a distraction.*

When we returned to Española, the men who had been
marooned with me at the Bay of Santa Gloria continued to swim.
Soon others took up this pastime, and it threatened to become
quite distracting for the populace.

I was no longer governor of the colony, but nonetheless
ordered all the citizenry to desist, to resist the temptation, as I
was resisting. But to no avail. Perhaps my role in history was to
forbid the practice of natación, and thereby ensure its popularity
amongst a rebellious populace, prepared to go to any lengths to
ignore my authority. I am a man more swum against than swimming.

That I do not recognize which of my discoveries are the most important and that I concentrate too much on aspects which are fickle and fleeting. That I pester Their Majesties with entreaties for redress and reinstatement.

There are people who think this because they hear only that I hire lawyers and seek redress for fiscal wrongs.

Remember: Don Fernando and Donna Isabella, by the Grace of God King and Queen of Castile, Leon, Aragon, Sicily, Granada, Toledo, Valencia, Galicia, Majorca, Sevilla, Sardinia, Cordova, Corsica, Murcia, Jaen, the Algarves, Algeciras, Gibraltar, and the Islands of Canary, Count and Countess of Barcelona, and Lords of Biscay and Molina, Dukes of Athens and Neopatria, Counts of Roussillon and of Cerdagne, Marquises of Oristano and of Goziano, counting the many good and loyal and grand and continuous services which I, their servant have done them, particularly in discovering and placing under their power and sovereignty the islands and continental land which I have discovered in the Ocean Sea, empowered me in my office of Admiral of said Ocean Sea, which is of the part to the west of an imaginary line which his Highness ordered drawn from pole to pole 100 leagues beyond the Azores and Cape Verde Islands, beyond which they ordered and made me the Admiral of the Sea, with all the pre-eminencies due an Admiral; and they made me their Viceroy and perpetual Governor for ever after, in all the islands and main lands discovered and to be discovered, for myself and my heirs.

Their Majesties do not empower lightly. I seek no more than this due, more for my children than for myself.

Father Gorricio, who is with me today in my chambers transcribing my dictation, remarks 'Anyone who remembers all the titles and honours of Their Majesties, and further remembers the order of importance of those titles, without the use of a lawyer, such a man deserves what is due him. '

That Isabella judged the High Voyage a failure and was in the end disillusioned by all I had wrought.

I saw the Queen for the last time in the Royal Library in Segovia. It was November 25th, 1504, two weeks after my return. She was not well and had had her bed moved into the library, midway between the fire and the window, which as usual looked out on her winter garden. The Royal Gardeners had attempted to plant here some of the strange flowers, shrubberies and fruits from the islands of the new world, without success. Beside her was her bookstand, this time filled with note pages.

'We have failed, Admiral. Both of us. I have stopped reading travel tales and have begun to write my own. Imaginary, of course, but entertaining nonetheless. And you have returned to stay.'

On her bedcover, which was fashioned from the Royal Ensign with the letters F and Y above the crests, I laid my gift. It was a crucifix from the church in the settlement of Isabella on Española, made from the gold of the Indies.

'And this is a relic of our failure. We have taken the New World and made it old.'

'I have claimed these new lands in your name, Isabella. I have named cities, rivers, islands and continents after you and your saints. These glories will live one hundred and fifty years, until the end of time.'

'Even if they live five hundred years, Admiral-who-is-no-more, there will be no glory.'

She took the cross and kissed it. 'I have one more benefice to confer, Admiral. Now that I am going, you may stay. Stay and write your tale.'

The next day the Queen died, and I left for Valladolid.

Charges Concerning the Current Period of My Life, and My Divine Mission

That the King and Queen sinned against the natural order in elevating me to my present stature.

My brother Bartolomeo has reminded me that I am the first man to be born into a lowly guild family and now be great – a Viceroy, an Admiral, et cetera – and to have so risen only through the force of my vision and accomplishments.

Before me, the only paths for great men born poor were those of scholar or priest. I have shown that a man in his own lifetime can achieve a position previously gained only by birthright.

Possibly some of my supposed clumsiness as a governor and administrator may be attributed to this: All other such lords grew into manhood watching their fathers perform these tasks, they had teachers, they shared in such responsibility before assuming it all. My experience is as a captain of ships and fleets. Few men have experience in successfully governing greedy settlers. There are no useful apprenticeships to be had.

Nonetheless, I still have my titles. The King and Queen were sufficiently old-fashioned not to have removed them under pressure from my fellow nobles.

Perhaps Their Majesties' most dangerous action was in recognizing that ideas can overcome blood. If this is true and unless this notion be exorcised, then my real accomplishment may be in toppling the power invested through birth alone in lords of the realm.

There is an irony to be savoured, not just that the son of a woolgatherer should become an Admiral and his sons after him be Admirals too, but that he should be the first to recognize the profound changes implied by the fact that the son of a wool-gatherer should become an Admiral and his sons after him be Admirals too.

*That I have placed absurd and vain obligations upon
my descendants.*

I have obligated them with only two charges, that they shall distribute in accordance with certain guidelines the revenues of the trusts I have established for them, and that they shall bear my arms, without inserting anything more in them, and shall seal all correspondence with the seal of these arms. They shall sign with my signature which I now use, which is an X with an S over it and an M with a Roman A over it, and over that an S and after it a Greek Y with an S over it, with its lines and points as is now my custom. And they shall write beneath only 'The Admiral', although they may be given and may acquire other titles from the King and his descendants. In their own enumeration of titles, they will be able to write all their titles as shall please them; solely in the signature will they write 'The Admiral'.

Other fathers leave their children far more complexity than this. Some leave their children poverty, or fear, or membership in a guild of criminals.

That I have been purposely obscure when asked about my origins, refusing to be precise in answers to such simple questions as when and where I was born.

I have been asked these questions and I have answered them.

Of what advantage would it be to me to claim a wool-weaver for a father if I indeed had descended from nobility? To claim kinship with pirates? To be older than I am?

That I should have stayed a wool-comber as my father
and his before him.

There may be a moment in one's life when one can choose a path
with no tragedy, and not attend to the voices of saints in one's
ears. One can stay a wool-comber.

I was born in Genoa at the Porta Sant'Andrea in the year
1446. As a child I wandered amongst strange ships on the Mole
Vecchio and listened to strange tongues talking of the terrors
beyond the seas.

Each day my mother would take me to the churches, each
built with the black and white marble that I came to associate
with all opportunities and all choices. Santo Stefano, home of the
first martyr, a Jew stoned to his death for calling other Jews
stiff-necked and uncircumcised. San Matteo, the church of commerce of the D'Oria family. San Lorenzo, where lies displayed
the Sacro Catino, the glass and emerald cup of Our Lord's final
supper, brought from Caesarea, object of vain worship. Here was
the birthplace of my vision of a New Crusade financed by the as
yet unattainable riches of the lands beyond Jipangu. San Siro,
Church of the Apostles, where I watched odd men in odd hats
elected as the Doges of Genoa.

I was then put by my father into apprenticeship in the Secret
Guild of Wool-combers, not so much secret as a guild of difficult
work and worn, tired hands. At fourteen these hands were
already apt in designing the sphere of Our Lord's Creation, and
placing on it the cities, mountains, rivers, islands and ocean seas.
These eyes had read much.

But then picklock primes seconds abb and breech. Remove
the kemp, burn the bad wool, take in the ammonia to deep
lungs. Picklock primes seconds. The rhythm of separate, card and
comb, separate, card and comb. Seconds abb and breech. The
rhythm of waves in the nearby harbour, waves in the sea. Picklock primes and heads. Aristotle. Caesar. Pliny. Turn the wool.
Turn the page. Ahmed Ben Kothair. Ibn Rashid. Rabbi Samuel de

Israel. Seconds, abb and breech. Joachim di Fiore. Duns Scotus. The Imago Mundi of Petrus Aliacus. Wool to cloth. Ships to sea. The Travels of Signor Milioni. From the softest shearings of merino lambs to the bristles of the wild boar.

With twin views of the world, I lived among the black and white marble of the churches. One view created cloth and wealth, with the other I wrote two books. My *Notebook of authorities, statements, opinions and prophecies,* and *The Five Hospitable Zones of the Earth.* The one leads to the Final Crusade, the other to the lands beyond the Ocean Sea.

Here Jerusalem is free. Here intercourse with the Indies thrives. My books change history: My history as a wool-comber is without change and peaceful.

I am before the Grail in the Church of San Lorenzo, and like the city of my birth, I am Genu, Genoa, on my knees.

That I should have remained a cartographer, a calling
in which I have some skill.

There are but few higher callings than the making of maps. Here it may be said that the pursuit of success in business matches the needs of science: to amass as much detail from as many sources as possible and draw the most accurate map, weighing the credibility of each source, the plausibility of what is reported, the consistency or inconsistency with what has been drawn before, and the credibility of those sources.

If one accomplishes all this, and becomes known for the accuracy and the credibility of one's maps, then one is able to construct a business life of drawing maps.

Each explorer whose voyages are described in a published book provides his piece of the map of our circle of the earth, and sheds light on his small corner.

Only the cartographer assimilates all the knowledge, and especially the most current knowledge from the most recent expeditions, and brings this together, into one picture. If the picture changes one month later, he must draw a new map as he knows his concurrents will.

The Pope, Emperors and Kings, merchants, tradesmen, and all the explorers who come afterwards turn to the same map as a foundation for their visions of empire, their trading plans, and their next voyages.

Without cartographers, there can be no progress. Without explorers, there can be no cartographers.

That I am a man of little loyalty, having sought support for my voyage from rulers in France, Portugal, Holland and Castile.

I am intensely loyal, immensely loyal, and would have felt the same toward any ruler so enlightened as to see the sacred and imperative nature of my vision. Yes, I knocked on doors as a gypsy or tinker might, house to house, but I knocked only on the best doors.

There was no prophecy to indicate under whose banner I should sail. I had no instructions. Flags meant little to me, and I admit they still do. I admit now too, for example, that I do not understand the fuss over the French maiden, Johanna la Romée of Domremy, who led the French in battle against the English. How can a battle between Christian powers further God's plans? The only earthly wars in which He has a stake are those of Christian against infidel.

I take His side. I was prepared to sail in 1492 and I am prepared to sail again today under any Christian flag that recognizes the holy nature of my mission.

I am loyal to Their Catholic Majesties, Ferdinand and Isabella, insofar as they recognize and perform God's will. In the fullness of time, like a gypsy, God has no interest in flags and boundaries. Unlike the much-slandered gypsy, God takes what is rightfully His.

I slander them too in my metaphors. A cheap jibe at their expense.

*That I grow old in obscurity and that my discovery
of the Indies is a sham.*

I am not rich, but I have seven attendants and a stream of visitors
and petitioners. I have not received my due income from the
Indies, but I will leave much gold to my family, to acquaintances
in Genoa and elsewhere, and will still have small amounts left
over to bequeath anonymously to certain individuals who have
shown me kindnesses in my life.

The Indies can be reckoned by no measure to be a sham or
delusion. Since my first voyage, hundreds of voyages, carrying
thousands of men, have crossed the Ocean Sea, transporting
wealth in both directions, and carrying the word of Our Lord to
attendant heathens.

That I am jealous of Americus Vespucius for the ease with
which he obtained support for his voyages and for his
commercial success.

His Highness believes that the ships of Americus Vespucius went
to the best and richest of the Indies, but those labours have not
been of the profit that they should have been; and fortune has
been adverse to him as it has been to me.

Americus Vespucius is a good man, and a man to whom I have
entrusted secrets of my affairs. He is always desirous of pleasing
me, and is acting to accomplish certain actions in my interest, if
he is able. I am pleased to be able to provide him with advice in
matters of navigation and of the commercial aspects of the Indies,
in which pursuit he is now engaged on behalf of Spain.

This explorer is much travelled and has seen more of the New
Lands than any other man, travelling south several thousands of
leagues from the Great River of Eden that I discovered. He is
skilled in all navigational techniques and will provide great service to the Royal Court.

But for myself, there is no finer sailor than Vespucius. I trust in
history, and history will give him his due.

*That to ensure recognition of my place in history, I must write a
full account of my expeditions. That in the absence of such works,
others, less noble than I, who contribute much less than I to the
accumulation of knowledge, will, nonetheless, put themselves
forward self-importantly as explorers and discoverers.*

There is nothing left to discover, and little time left to make such
explorations. I have written in my *Notebook of authorities, statements, opinions and prophecies* (which my son Ferdinand calls my
Book of Prophecies) all that is to unfold until Our Lord comes
again.

There is no reason for me to be concerned about history
recognizing the importance of my true role. I have done my part.
I have set the steps in motion. History is over.

*That there is confusion over the authority ascribed by me
to the Prophecies of Esdras.*

The Holy Word of Esdras is clear on this matter. In the sixth chapter of his fourth book, he tells us that on the third day of Creation, God commanded that all the waters of the Sphere of our Earth would be gathered up in the seventh part of the earth, and the other six parts would become the dry land we know now as continents.

Such a calculation is important because it helped me to recognize, when I was sailing in the south of the Ocean Sea, that I had sailed one seventh of the circumference of the earth and had in fact reached the northern edge of the Great Garden in which God created us.

A man can enjoy no higher joy than that of seeing, with his own eyes, the great fresh-water rivers of Eden flowing into the Sea. I am not sufficiently wise to know whether the tangle of waterways that we saw at Boca del Drago was of one river or of all four, but these waters were of such beauty, such force, and such majesty, that none who saw them were not moved to tears and to prayer.

Knowing that God had sent Adam and Eve out from this place, naturally, we feared to make a landing party. Susana Fontanarossa was curious to see the tree at the centre of the garden and to converse with the Serpent, who we all agreed must be immortal and live still. Still, we became anxious, and sailed on quickly.

That I frighten the unschooled with my predictions.

The Sacred Writing testifies in the Old Testament by the mouth of the Prophets and in the New Testament by our Redeemer, Jesus Christ, that this world will come to an end. The signs of the time when this is to take place Saint Matthew tells and Saint Mark and Saint Luke. The Prophets also have abundantly predicted it.

Our Redeemer said that before the consummation of the world, all that was written by the Prophets will be fulfilled.

Saint Augustine says that the end of the world is to come in the seventh millenary from its creation. The sacred theologians follow him, especially Petrus Cardinal Aliacus in Word XI and in other places, as I will tell below. From the creation of the world, or from Adam, until the coming of our Lord Jesus Christ there are 5345 years and 318 days, according to the account of the King Don Alonso which is considered the most certain. Cardinal Aliacus, in Word X of *Elucidario Astonomice Concordie cum Theologica & Hystorica Veritate* , adding to this 1501 years, makes in all 6845 years. According to this account, there were at that time only lacking 155 years to complete the 7000, in which year I said above, according to the said authorities, the world must end.

But why should such talk frighten anyone? The End of this World is the beginning of the next. We who have served in its manifestation are the foundation of the next. We carry Our Lord in our hearts, and on our backs.

I appreciate the irony of my position. Unlike many who think their art, their inventions and their written wisdom will live forever, I leave books to a posterity I know will only last for 150 years. Is an accomplishment less worth doing if it outlives its creator by 150 years instead of seven millennia?

That I am obsessed by the idea of the End of the World.

Compared with this one, what other obsession is worth having?

Anyone working in the service of the Lord can have no higher ideal than to prepare the way for the Second Coming of the Christ. All my actions, my arguments, my decisions and my voyages were made for the greater glory of God in fulfilment of the prophecies that all the world should be saved before Jesus, His Son, shall come again.

When I forgot myself and forgot this ideal and allowed myself to believe my personal goals were more important than God's, then God, in His wisdom and mercy, saw fit to punish me. I take some comfort from the fact that He also punished my enemies, sometimes unto death.

Now the Missionaries must do their part. It is not up to me alone to accomplish the Building of the Foundation of the End of the World.

That I am obsessed by hats.

Curious that Father Gorricio should remind me of this charge. I have not thought about hats for several years.

M. Polo describes the headgear worn by the inhabitants of Cathay and India – headgear both practical and distinctive. When first we landed at the islands of the Indies, I did not hide my dismay at seeing the natives all bare-headed and at recognizing thereby that we had not yet met the people described by M. Polo in his texts.

A sailor on the first voyage, a man who had lived in Roma, to celebrate the Feste de la Bufana in his family's traditional manner, fashioned himself a cloth hat which he intended to wear for the three days of the festival, to honour his family and his town. Someone of his friends remarked that this action would make me happy since the hat was most probably very like the hats I sought.

Soon the plans became more elaborate. Five of my men made similar hats and a sixth was appointed to lead me to a part of the forest at dusk where the five would disport themselves in the shadows. My guide and I would watch from such a distance that I should not recognize the sailors.

Here the story was made more complicated by events. The men decided to celebrate Bufana in the traditional way, with banging of drums and metal items, and the making of much noise. The din attracted a small group of natives who felt the gathering to be very invigorating, and soon joined in. Moved by the spirit of the occasion, never having had direct content with the Indians before, my men placed their ceremonial hats on the heads of several of the natives there.

By the time I arrived, refusing to stand at a distance, to the consternation of my guide, several of the men were hoarse from the merriment. They rallied, however, to point to the natives' headgear and to shout, 'The hats! The hats!'

'These are wonderful hats,' I said, 'as good as any described by Signor Polo, and made carefully, of the finest Spanish cloth.'

*That I announced in 1501 that I was the Messiah anticipated by
Joachim di Fiore.*

This is an exaggeration.

That on occasion my behaviour may have been a source of confusion.

Two old men sat in the garden of my home in Valladolid in the autumn of 1501. The shaman of the island that we call Española, Palahuelo, also called Kee, had journeyed to Sevilla where he attended the Royal Court and asked Their Majesties to stop sending ships and colonists to the Indies.

His argument was very simple, and he repeated it for me one evening. I had not been invited to take part in his meeting with Their Majesties although to a great extent I was the subject of discussion there.

'I asked them,' Kee said, 'how they could be both gods and men at once. I told them their behaviour was inexplicable, and that I wished my people to be spared the confusion and embarrassment of having to live with them.'

Naturally, I expressed curiosity as to his meaning.

'The confusion began with you,' Kee said. 'You arrived dramatically on our shores, with the tallest vessels we had even seen, the loudest noise-makers, the most unusual colours of hair and costume.

'You presented yourselves to us as if you were omnipotent – even if unpredictable – and capable of single-minded pursuit of what appeared to us to be arbitrary and flexible goals. I see now that you pattern yourselves after your own God and His Prophets and Saints, and that you are not gods, but ten years ago I was not so well informed.

'But recall what happened next. A man and his crew were cast ashore on a beach, ill, weary, but still inconceivably helpless, unable to catch their own fish or gather food from an island known for its abundance.

'My brothers who live on that island were sympathetic. They knew of you. They felt perhaps it was their duty to help these men-who-might-be-gods. Imagine their surprise when they discover these creatures are fighting amongst themselves. Imagine

154

their horror when they offer to help some of your men canoe across the channel to Española and their men are killed by your mutinous crew. And then imagine their feelings. You call them together, tell them they must continue to bring you food or you will eliminate the moon, and that you will prove the strength of your God that very night.

'And then you do. A man practically too sick to get off his bed of palm leaves, who cannot convince his men to catch fish or gather fruit, this same man hides the moon on a cloudless night! Such a god, such a confusing, irrational religion that gives such authority to such a god and provides such miracles and such weakness.'

'How did Her Majesty respond to your request?'

'I like her,' said Kee. 'I, like her, can imagine a version of our history in which she sails with you on your first voyage, sees in my people the character which she says she admired in me today, offers to marry her daughter to one of our Chiefs, and in this way concludes the traditional merging of empires associated with her blood and ancestry.

'I cannot blame you that she did not sail with you, nor can I blame you for not recognizing that my people would have preferred marriage to massacre.

'The Queen told me that when she was five years old, her grandmother wanted to give her a present and asked her mother what the child would like.

'The mother was very fond of the garden and had recently become swept up by the popular idea of taking garden plants and growing them indoors. She told her mother, Isabella's grandmother, that the young princess very much wanted an araucaria plant in a wooden box. The grandmother found one, and although she remained convinced that plants belonged outdoors, for her granddaughter, was willing to present it in a box.

'The child Isabella, of course, had no interest in this plant, and was very disappointed in the gift. Her mother was only too pleased to look after it. Since that time, in the Queen's family, any gift designed to give pleasure to those who give instead of receive has been called an ''araucaria''.

'Just before the end of our interview, Her Majesty asked her maid to go quickly to the Royal Chambers. She returned with a small package wrapped in a silk cloth.'

Kee reached into a pocket in the fold of his clothes, and placed the package on my table. He untied the knot to reveal a small cutting from the Queen's araucaria plant.

He said: '''This is our civilization,'' she told me, ''and our religion, and our science. It seems to do well in most soils, and most climates, whether you water it or not.'''

That, although I acted as an agent of change, I failed to recognize the speed of those changes.

I am caught between a belief that all change is so gradual that one cannot recognize the moment when one took for granted the new fact or the new point of view or the new sense of the shape of the world; and an intuition that all change is sudden and that our views of the world and how it works change in fits and leaps: One morning you wake up and books that formerly were only found in monasteries may be purchased from a bookseller, and everything changes. One day you meet a neighbour in the street who tells you that Granada is Christian again, and history turns a corner. One morning you wake up and there are inhabited islands between Christendom and Cathay.

Are these revelations any less of a surprise if one is an explorer, and thereby, one imagines, predisposed towards an expectation of the new?

No.

Would I have been less surprised to awaken one morning in front of a great harbour, with a Khan-bearing barge rowing out to greet me?

Yes.

Susana Fontanarossa tells me I need to have a set of rules to guide me in combining what I knew yesterday with what I know today.

'There are only so many combinations of possibility,' she tells me. 'If on the morrow after your first discovery of San Salvador, you had sailed into the Khan's great harbour, you would have been surprised at such raw citizens so near to the seat of power and trading.

'If you had sailed for months afterwards seeing no other islands and no other natives, the one island would have become such an object of curiosity that you would have spent weeks trying to find it again – a speck on the sea with no neighbours as landmarks.

'But instead, the worst happened. Over the next weeks you found other islands, all a little bit different, but for the most part, the same. And natives who were much the same as previous, as like you as one sailor is like another. Except for the cannibals, of course.

'So the surprise vanished. Islands with natives and islands without natives became commonplace. You wanted something special. Gold would have been good, or silver.

'There had been a moment when a great change in your world occurred. Now that moment became duller in your memory, much as a wave turns to a ripple and then to flat sea.'

She was right, but has become wrong. There was a second moment, when we returned to Spain, when suddenly all of Christendom experienced that monumental change in its view of the world.

Certainly, by the second day, the wave became a ripple, and then flat. Maps were redrawn, instantly, as if there had always been lands at such and such a distance west of Gibraltar, of such and such a shape.

But that moment has been relived, in argument and tale, by me in my *Journal* and published letters, and elsewhere. The suddenness of the moment has been preserved for all time, and therefore we know now that change can be sudden if recorded properly.

That 'to explore' is a verb with lost meaning; that exploration is an act of interference, of defilement, of transformation, like alchemy.

I have told already of our decision not to land, but to sail quickly from the mouths of the rivers of the Garden of Eden. We were all fearful, recognizing that one could not tread lightly on such shores, but that there would be consequences.

('Should the Pope be the first man to step ashore at the Holy Garden of Eden?' was one of the questions we asked ourselves.)

Susana Fontanarossa regretted our haste, reminding me that God brought us to this coast, showed us good anchorage, and that He judges bravery favourably, having invented it to good use.

I could only laugh, imagining Susana as mother Eve explaining to an earlier Adam that God set a certain tree in our midst that we might enjoy of its fruit, and that He judges bravery favourably, having invented it.

She remarked on my use of the word 'defile'. 'Curious that you feel we would defile this coast by landing, but not so many others.'

'We read these coastlines like books,' I said 'and occasionally scribble in the margins, by founding settlements or planting fields. We almost never know what lies beyond the first mountain ridge, whether continent or another coast. We live only at the edge, touching the thinnest surface of the lives of these lands.'

'For an arrogant man, you do yourself a disservice by understating the impact of your presence on this coast.

'Imagine that you never landed, but only found anchorages along these coasts and rested there, a floating trading post for the local Indians. Do you believe for an instant that such a presence does not defile? (I ask in this manner only because I believe you think that a greater presence does so.)

'If we have defiled any lands by our presence, then we have defiled this one by finding it. By recognizing that men in ships can reach it. That men less reverent than we might eat its fruit, might cut its trees, might seek its snake.'

That I do not understand miracles and that I do not understand the value of allegory.

My other brother, a village priest in Tardonancia, undertook a pilgrimage to Rome when he was still a very young man.

On this day under examination, he walked through the city towards the Basilica of San Pietro and noted a sign as on a shop's door that indicated the sale of Holy Relics. He lived in a very poor village, and his neighbours were like to work very hard, and when he saw this sign he thought that perhaps his small village church would benefit from having such a relic, that such a relic would call blessings upon his church.

He examined small fragments of bones and crosses in wooden boxes, and pieces of cloth, and collections of hair. He asked the shopkeeper the price of a relic of San Januarius, and another of Santa Aquilina. He had no money: he could afford neither.

Disappointed, he prepared to leave the shop, when the shopkeeper offered a small relic box, and the use of his tools. My brother took the knife and cut a lock of his hair, on the advice of the shopkeeper stepped on it on the dusty floor with his boot, and put it in the box. The shopkeeper asked for the name of a saint, and my brother told him San Tomas. The shopkeeper wrote the name on the small cloth strip he put in the box.

Out in the street again, my brother was astonished, even frightened. He put the box in his bag, at the very bottom of his bag, wrapped in a cloth, and walked quickly to the Basilica.

He prayed vividly and wildly, asking for direction, but not knowing the reason, not feeling that he had anything to ask, but desperate. Finally he felt he was ready to leave the Basilica and leave Rome to travel home. By the large door, he saw a small sign with an arrow pointing to a small room. The sign indicated that here relics would be blessed.

As if in a dream, my brother entered the room, and found the small box in his bag. Gently cupping this box in his hands, he approached the old priest sitting in front of him, who said not to be afraid. At that moment he recognized the Holy Father himself. He was filled with fear, and ready to turn to run out of the small room, and the Father said again, do not be afraid. Give me this box.

And then the Holy Father said, 'This is wonderful. This is your own hair, I can see it. You have made the people of your Church very happy.'

And he proceeded to perform a blessing, and to speak other words of kindness to my brother.

When my brother returned home, he put this box, still wrapped in a cloth, into the bottom of a closet, and mentioned it to no one. He left it there for several years until the long illness of our sister, who lived also in this village. She asked him one day, from her bed, after many months of illness, if he didn't have some remembrance of his meeting with the Pope that she could hold for her prayers, that would comfort her in her weakness and desperation. She insisted.

He told her about the little box with the hair of San Tomas in it, blessed by the Holy Father in the small room at the exit from the Basilica of San Pietro, and he placed it in her hands. She asked for the rites, and she smiled and said her pain was gone, and she died.

And as for the charge that I do not understand the lessons of allegory and that I am too firmly implanted in the ground, I have a simple response:

I have no sister. I have no brother who is a priest. There was no miracle at Tardonancia.

*That there is no servant girl, Susana Fontanarossa, that I have
fabricated her for this narrative or else imagined her on
my voyages.*

My brother Bartolomeo is conspiring with other parties and has
told Father Gorricio that the anguish of my present state is
increased by my continuous imaginary conversation with the
Queen's daughter's servant girl, Susana Fontanarossa.

This is madness. The records show – surely! – the presence of
a girl of sixteen years on the dock at Palos the morning of August
3, 1492, before we sailed. The girl who looked like our mother,
and had her name.

Bartolomeo has read these notes and asks me to record in
these pages that despite his presence on two of my four voyages,
he never met this girl.

I say he sailed on the *Santa Clara*, she on the *Niña*.

He asks if it is not true that she sailed with me on the fourth
voyage, and I respond affirmatively. 'Yet during your year of
shipwrecked peril in Yemyaica, she appears in no journals,' he
says.

She disembarked before then, I tell him. She would not have
survived a year on a beach alone with 200 men.

'But how could she know this fate awaited you?' he asks.
'How did she know to disembark?'

It was happy circumstance, I respond, she was always blessed
with luck. She prayed for us. Even I, who am not a man given to
mystical leanings, realized she was praying for our safety and sur-
vival each day we were gone.

Just as I pray for hers.

*That Susana Fontanarossa remained in the Queen's employ
through all my voyages, reporting to the Queen frequently, and
being the source of Her Majesty's otherwise inexplicable
understanding of my affairs.*

'Why are you here?' I asked Susana Fontanarossa as she visited
me in the middle of my legal battles with the King.

'Think of me as an angel, watching over you as protectress
and muse.'

'You come from God?'

'No. Think of me as Johanna d'Arc, leading your force into
certain battles and uncertain places. Think of me as a daughter
you don't know you have. Or a son, if you prefer.'

That if, before I began, instead of praying for strength, I had prayed for an understanding of exploration, and of all I was about to undertake, I would not have sailed.

Susana Fontanarossa had not read these pages when she made this accusation.

Only she could concoct such a charge. Naturally I asked her for an explanation.

'I have a theory,' she said, 'that our minds cannot comprehend the magnitude of what you and we have accomplished, and that neither our minds nor those of our children will ever come to comprehend that perhaps we should not have done what we have.'

'What we have done is to be curious,' I answered, 'but no more curious than the newborn who wants to leave the womb, or the child who must open her mother's cupboard door, or the young man who leaves his father's village. We have gone places and named what we saw, as did Adam and Eve.'

'Admiral, I am not Eve.'

'Nor I Adam. But you, and others who contemplate what I have completed, seek always to name my acts, then to lay blame, then to see what I have done as a personal injury, as part of your history as if you were islands, continents, tribes, whole races with no other chapters to your history than before me and after me, a poor wool-comber from Genoa with three ships.'

'Admiral, you exaggerate.'

'I hope so. I hope so. Then there will be no need to finish this book.'

'And no need to finish this conversation.'

*That even now, in this present work of matters so particular
and so personal, I withhold truth.*

I have a mission of which I have written in no other book, of
which I have spoken to no one alive.

The Holy Father, the Pope, never speaks of what I shall reveal
here; nor do the priests. There are two religions in Christendom,
and the Pope leads only one. There is another, and I have
difficulty speaking of it, although I am persuaded of its importance.

The most part of the priests pay scant attention and many of
the Royal Court feel we live in a time after the time of prophets
and saints. They believe that only the fathers of the Church of old
spoke with God. They do not recognize all that must be done
today, and the men and women who work to accomplish it.

In the Gospel of the blessèd Saint Mark, Our Lord instructs
His Apostles to go into all the world and preach His Gospel to all
creation. In the book of visions of Saint John, he described the
Second Coming of Our Lord and the Chosen, clothed in white
robes, whose number come from all nations, and peoples and
tongues. For this to come to pass, then each must first choose to
follow Christ. Our Holy Mother the Church accepts this and
sends Missionaries unto all heathens. And she makes plans for the
recapture of the Holy City of Jerusalem.

The Franciscan Juan Rupescissa has written that the conversion of the Tatars shall be followed by the conversion of the Jews
and the extermination of the followers of Mahomet. When this
comes to pass, there will be little left to accomplish. The only
event left in history will be the Second Coming of Our Lord, the
End of this World, and the Birth of the New Jerusalem.

I say that he rightly predicts the Final Days, but that the Holy
Spirit works in Christians, Jews, Moors and in all others of all
sects, and not only in the wise but in the ignorant: for in my time
I have seen a villager give a better account of the heaven and the
stars and their courses than others who expended money in

learning of them. And I say that not only does the Holy Spirit reveal the things of the future to rational creatures, but shows them to them by the signs of the heavens, by the air, by the beasts, when it pleases Him, as it was with the ox which spoke in Rome at the time of Julius Caesar, and in many other manners, but very well known to all the world.

Such prophecy continues to this day, and the Holy Spirit reveals His truth to divers men and women who are not among the priests and bishops. These truths, then, sometimes coming from the unschooled, are not always recognized as truth by the priests.

By leading the armies of Their Holy Majesties into battle with the banner of the Pope, I will bring to consummation the prophecies of those who otherwise are unduly shunned by Our Holy Mother the Church.

I take comfort from my knowledge that, traditionally, a prophet's life should be difficult. Otherwise one might be suspicious of his motives.

*That no man has suffered such ignominy as that implied by the
collection of accusations here gathered together. That I have
invented some or many of these accusations in order only that I
might refute them, that being accused of so much is a kind of
flattery for me.*

I have been accused of sins and crimes beyond all understanding.
It would be an act of extraordinary fantasy to invent charges as
fantastical as the truth.

On my second voyage, I was accused of humourlessness. This
because I didn't laugh with some of the men at one of their jests.

The loudest one told a tale of three Portuguese fishermen
from the village of Cascais. The men of the village shared a set of
some fifteen small boats for fishing, and on a certain day three
friends went at dawn to fish. Within a very short time, one of
them died, and the others threw his corpse into the sea. A few
moments later, one of the men realized that the man's wife
would be angry if she couldn't bring flowers to put with the
corpse, and he started to panic.

'Don't worry,' said the other, 'we'll put a stick of wood here
in the water to mark the place.'

'That won't work,' said the other, 'there is a wind, and the
stick will float away.'

'I have a piece of chalk-stone here,' said the first fisherman.
'We'll just put a cross on the side of this boat.'

'This is not a bad plan,' said the other, 'but how can we be
sure to get the same boat tomorrow?'

I didn't laugh when I heard this jest, and from that day forward
was labelled as a man of no humour.

How could I laugh at such a story? I heard it first from my
grandfather, when I was ten years old, but it was about fishermen
from Corsica. The farther one travels, the more one learns.

That I ascribe poetry to situations in which there was no poetry.

Bartolomeo, my brave and cynical brother, has read some of these passages.

'The Indian Palahuelo, about whom you have written,' he said to me, 'was neither as eloquent nor as poetic as he seems from your pages. Others who knew this man found him abrasive and crude.'

'Others?' I asked.

'Me. Others.'

'The Queen,' I told him, 'a wise woman, full of insight, told me on several occasions of her pleasure at meeting Kee, and corroborated his account of her statement that had she imagined such as he, the history of the Indies would have been different. She chastised me for bringing back only young Indians on my first voyage, in Kee's words, "so full of life and so empty of wonder, so unable to portray a rare and precious civilization by standing silently." She wished her daughter could have married his son.'

Bartolomeo was silent. Short bursts of eloquence, from anyone, often quieted him. I continued.

'I am no poet. I cannot and do not create poetry from the events around me, nor do I particularly look for the poetic moment amidst our hours, or the poetic view within this landscape.

'I tell you what I have seen and what I have heard. The last time Kee was with me, he left the araucaria plant that the Queen had given him, and said, "My friend, this is your present, as much as it is mine." This, Bartolomeo, is a poem.'

That I write like a man guilty, that I am too quick and too
elaborate in my defences, and that I therefore must be
guilty of all the charges against me.

I will outlive you all, my accusers.

History will remember me as a strong, brave man acting in the service of his God.

In my small way, I contributed what I could, and my contribution was great. I did no more and no less than what would be expected of any loyal subject of Their Majesties, of any man wise enough to assemble the facts and the clues, and of any Christian.

I accept the blame for my shortcomings and the responsibility for bringing Christendom into contact with the old world across the Ocean Sea (which we now call the New World), for proving the circumnavigability of the globe, and for offering to all the world the lessons and joys of the Christian faith.

History will remember me as acting in the service of currents stronger than I.

That I failed in My Mission.

If any feel that I have not accomplished the Mission assigned me by God, they have only to examine the facts, to secure the list of other explorers and missionaries who continue my work, who follow the easterly and westerly sailing routes established by me.

I also felt as you do, that the True Sign of the success of My Mission would be the completion of the prophecies and the return to this life of Our Lord. I have been as guilty as many of the Apostles and Saints before me in believing that I would be the one to witness the Accomplishment of His Purpose with these eyes. It is not for us to know the timing of God's Divine Plan, only to perform our part in it.

I was called, in my weakness, to accomplish the prophecy, and in my pride I imagined, indeed prayed, that I would be allowed to witness His Triumphant Return.

My friend Father Gorricio suggests this is not the statement of a humble man. 'I have been punished already for my vanity,' I reply, 'and having paid so dearly, I would be foolish to become humble so late in my life.'

My punishment is God's recognition of the importance of my work. There are those who say that I was a wealthy man, that my first voyages brought me significance, and that I have offended God, and have been accordingly brought to my present situation.

They do not understand His Ways. He spoke to me in my garden, when I had spent three days in penitence and fasting, and He showed me that I must find His Garden and the Great Rivers of Eden, and He showed me that I must sail further to the South, and He told me that if I were successful in this Mission I would be blessed with poverty and weakness and that I, alone among men, would recognize the Delay in the Fulfilment of His Holy Word as being, itself, the necessary hesitation, the moment that God gives us to baptize the heathen, to confess our lives, to prepare for that for which we can never prepare.

Like you, I thought that the World would End in Fulfilment of the Prophecies each time I returned from a voyage. I see now that there is much work to be done to prepare ourselves to receive Him, and I no longer expect to witness this moment from Earth.

But do not imagine that I am jealous of you who will suffer the Extraordinary Manifestations of the Last Days. I will be at the Right Hand of Jesus, and I will intercede on behalf of all who have shown faith in me.

I say this with all appropriate humility, trusting that, as I have heard God's voice and heeded it He will once, perhaps, listen to mine, and that I may, on the strength of my Service to Him, commend you to Him.

I have answered you all, my accusers, and all your accusations. I welcome you to the era foreseen by the Saints, in whose service I acted in fulfilling my part of their prophecies. I have brought you to a New World, and I forgive you all.

A Few Words of Epilogue

Reader: Susana Fontanarossa has now read these pages and made the accusation that I do not understand my place in history.

We were standing together, in front of the fourth station, in a small garden near Valladolid. The garden surrounds a chapel dedicated to Santa Maria del Giglio on the grounds of the summer estate of a wealthy Venetian merchant from Cadiz. The garden is full of lilies and hedges, which form a maze whose each turn encloses a station of the cross. Following the habit of my brothers the Franciscans, we were meditating and praying at each station. In between, as we decided which path to follow, we continued our conversation about history. (As I grow old, all my conversations are about history or money.)

'I am but a boy whose mind was filled with sailors' tales of adventure, my eyes with a brother's maps.' We were at the seventh station, depicting Jesus' second fall on his passage to the Crucifixion.

'There have been many such boys, Admiral, but in the future their minds will be filled with your tales, and the maps have been changed forever. No, childhood imagination is no defence. It is true that childhood is overlooked by historians. And people who make accusations of you have little imagination.' We passed the eighth station.

'But you understand? You have had your dreams as a girl?'

'I am still a girl and have many dreams. That is why I am here with you. But I have something more. I have the perfect defence against all accusations.'

'That I will answer to God?'

'That is much too common a thought. But perhaps in my two points one could find God.' We turned and were before the eleventh station, the scene of the Crucifixion.

'Time and history,' said Susana Fontanarossa. 'You see, there was a time before you, Admiral, when the islands you discovered were filled with gentle people and flowers like

173

these.' She picked a lily and crushed it between two fingers. 'Then someone came and discovered and killed and pillaged and despoiled the land. That someone was an ancestor of your friend Kee. An ancestor who then told a tale like yours. And before that, those gentle people came and discovered and killed and told that they were gentle before they were discovered and killed by Kee's ancestors. Even Kee and his people were being discovered and killed and eaten by the Caribs when you first arrived. Somewhere there is a tale full of woe accusing them. And somewhere, from long ago, are the tales of accusations against the Indians who discovered and killed and named a continent and islands after themselves. And the Indians were discovered and killed and their lands added to the maps by Alexander. His historians wrote books much like yours. As also did the Caesars. And the Prophets of the Jews. And Pharaoh. They all came and discovered and stand accused.'

We were now lost amidst the hedgerows.

'Then all history is but a denial of accusations, and I have done what many men have done.'

'You have done what every man has done.'

'Then I will have to answer to God.'

'Of course,' Susana said, 'but you should let time and history answer for you.'

'I cannot wait for them.'

'Then finish your book, old Admiral. More than most books of history, it has the charm of candour. Few other books proclaim so boldly that they consist of a denial of accusations.

'This book has a special place in my heart,' she continued, 'since I am in it. I suspect there will be numerous other books written of our voyages, but that only in these accounts do I play a part .'

We had become so sufficiently lost in the maze that we had failed to pass the twelfth station (or had not noticed it) and thereby avoided The Death of Our Lord.

Susana Fontanarossa mentioned that this garden was reputed to have fifteen stations, the last one being The Discovery of the Cross by the Blessèd Helena. She wished to know 'as a fellow discoverer' the truth of this and left me standing by the thirteenth station, meditating on The Descent.

Colophon

Christopher Columbus Answers All Charges was created from files conforming to the International Organization for Standardization's Standard Generalized Markup Language using *SoftQuad Author/Editor* wordprocessing software. The files were marked up in accordance with the International Committee for Accessible Document Design's (ICADD) Braille extensions to the SGML document type definition of the Association of American Publishers. The book is available in a large print edition from the publisher, in electronic form for persons with disabilities from Recording for the Blind, and in the Braille version from the Canadian National Institute for the Blind.